REMASTERING MUSIC AND CULTURAL HERITAGE

Remastering Music and Cultural Heritage presents a detailed account of the culture and practice of remastering music recordings. By investigating the production processes and the social, nostalgic and technological components of remastering practice, the book demonstrates the application of these techniques to iconic recordings by artists including The Beatles, Elton John and Oasis.

Through comprehensive interviews with music production professionals directly involved in both the original productions and remastered releases of these iconic recordings – and detailed digital audio analysis – this book offers an extensive insight into music production and remastering practice. Readers learn about the music production techniques behind creating some of the most well-recognised and loved albums of all time, as well as the processes used to create the remasters, to help guide their own projects.

Remastering Music and Cultural Heritage is essential reading for students and teachers of music production, professional practitioners and musicians.

Stephen Bruel is currently Programme Leader and Senior Lecturer for the Sound and Music Production degree at the University of Lincoln, UK. Stephen's published research includes remastering Australian bands Sunnyboys, The Lyres and his previous bands, Jumble Sale and Ben's Calf.

PERSPECTIVES ON MUSIC PRODUCTION

Series Editors: Russ Hepworth-Sawyer, *York St John University, UK*, **Jay Hodgson**, *Western University, Ontario, Canada*, and **Mark Marrington**, *York St John University, UK*

This series collects detailed and experientially informed considerations of record production from a multitude of perspectives, by authors working in a wide array of academic, creative and professional contexts. We solicit the perspectives of scholars of every disciplinary stripe, alongside recordists and recording musicians themselves, to provide a fully comprehensive analytic point-of-view on each component stage of music production. Each volume in the series thus focuses directly on a distinct stage of music production, from pre-production through recording (audio engineering), mixing, mastering, to marketing and promotions.

3-D Audio
Edited by Justin Paterson and Hyunkook Lee

Understanding Game Scoring
The Evolution of Compositional Practice for and through Gaming
Mackenzie Enns

Coproduction
Collaboration in Music Production
Robert Wilsmore and Christopher Johnson

Distortion in Music Production
The Soul of Sonics
Edited by Gary Bromham and Austin Moore

Reimagining Sample-based Hip Hop
Making Records within Records
Michail Exarchos

Remastering Music and Cultural Heritage
Case Studies from Iconic Original Recordings to Modern Remasters
Stephen Bruel

For more information about this series, please visit: www.routledge.com/Perspectives-on-Music-Production/book-series/POMP

REMASTERING MUSIC AND CULTURAL HERITAGE

Case Studies from Iconic Original Recordings to Modern Remasters

Stephen Bruel

LONDON AND NEW YORK

Designed cover image: Shutterstock

First published 2024
by Routledge
4 Park Square, Milton Park, Abingdon, Oxon OX14 4RN

and by Routledge
605 Third Avenue, New York, NY 10158

Routledge is an imprint of the Taylor & Francis Group, an informa business

British Library Cataloguing-in-Publication Data
A catalogue record for this book is available from the British Library

ISBN: 978-1-032-01230-8 (hbk)
ISBN: 978-1-032-01229-2 (pbk)
ISBN: 978-1-003-17776-0 (ebk)

DOI: 10.4324/9781003177760

Typeset in Times New Roman
by Taylor & Francis Books

CONTENTS

FIGURES

TABLES

ABBREVIATIONS

A&R	artists and repertoire
BTR	British Tape Recorder
DAT	digital audio tape
DAW	digital audio workstation
dB	decibels
DI	direct injection
DR	dynamic range
DSD	direct stream digital
DXD	digital eXtreme definition
EQ	equalisation
FFRR	full frequency recording range
FFT	fast fourier transform
IEC	International Electrotechnical Commission
IPS	inches per second
ISRC	International Standard Recording Code
ITU	International Telecommunications Union
LSO	London Symphony Orchestra
LUFS	loudness units relative to full scale
MFSL	Mobile Fidelity Sound Lab
PQ	production queue
RIAA	Recording Industry Association of America
RMS	root means squared
RT AVG	real time average
SACD	super audio CD
SPL	sound pressure level
SR	spectral recording
UPC	universal product code
WAV	waveform audio file format

ACKNOWLEDGMENTS

I would like to thank all of those people who supported and encouraged me through my journey of writing this book. And it was a journey!

In particular, a big thank you to Perspectives on Music Production series editor Russ Hepworth-Sawyer and publisher Hannah Rowe from Routledge for their support and guidance to not only give me the opportunity to write this book but confidence to finish it as well.

This book would not have been possible without the professionalism and enthusiasm shown from those who contributed their expertise, knowledge and time. I would therefore like to thank in chapter order Will Prentice and Mike Dutton who shared their comprehensive knowledge of audio restoration techniques and remastering from vintage audio formats. My sincere gratitude to Nick Webb, retired from Abbey Road Studios, for his detailed account of working with The Beatles and recording their *Abbey Road* album as well as mastering Oasis. A big thank you to Sam Okell for his thorough description of the remastering process used for the 2009 The Beatles' release, and to Beatles' assistant recording engineer John Kurlander for his thoughts.

I would like to thank well-renowned composer and recording engineer David Hentschel and mixing supremo Peter Kelsey for their invaluable contributions regarding recording and mixing Elton John's *Goodbye Yellow Brick Road.* Thank you also to Ray Staff for his portrayal of mastering this album and Tony Cousins for his comprehensive and technical description of the process for creating the 1995 remasters. A big thank you to music producer Nick Brine for his thorough and detailed account of recording and mixing Oasis' *(What's the Story) Morning Glory?* album and Ian Cooper for his thoughts around remastering this album in 2014. My sincere gratitude also to mastering and remastering engineer Paul Baily for his description of his work in the orchestral and classical fields. I would also like to thank all participants in my book for contributing their thoughts around remastering in general, and the potential impact of this practice on cultural heritage, nostalgia and authenticity.

Finally, and by no means least, I would like to thank my family and friends for all of the love and support they have given me during this endeavour. In particular, I would like to thank my wife and best friend Janice Holland who has been my rock and whose unconditional love made it possible for me to see this book through unto the end.

INTRODUCTION

Since the inception of the compact disc format, there has been a push by record companies to reissue digitally remastered versions of iconic studio albums in various digital formats (Rumsey, 2012). Albums originally released on vinyl from iconic artists The Beatles, The Rolling Stones, Elton John, The Who, Queen and Oasis have all been digitally remastered and re-released through their respective record companies (Benzine, 2006; Richardson, 1997; Vickers, 2010). This phenomenon has led to discussion and research into the elements of remastering, including obvious volume increases made possible through the technological aspects of the digital CD format (allowing a greater loudness level on playback) and the inherent diminished dynamic range as a result of increased loudness that has been well documented under the terms 'loudness wars' and 'hyper compression' (Hjortkjaer & Walther-Hansen, 2014; Nielsen & Lund, 1999; Vickers, 2010; Walsh, Stein & Jot, 2011).

Building upon my previous research into remastering practice (Bruel, 2019), in particular iconic Australian band Sunnyboys' eponymous 1981 debut vinyl release (Bruel, 2021a) and 'lo-fi' studio recordings of my previous bands Jumble Sale and Ben's Calf (Bruel, 2021b), this book further investigates the phenomenon of remastering. It begins with an analysis of what mastering and remastering involves, and its transformation primarily through technology from a purely technical process to a more creative and artistic activity. From there it examines the remastering process applied to vintage media storage including wax cylinders, shellac discs (acoustic and electrical), lacquers/acetates, tape and other various rare formats.

To identify sonic variances between an original release and its remaster, the book then explores the original production and mastering processes as well as remastering practice applied to a selection of iconic UK recordings emanating from different decades. These recordings consist of The Beatles' *Abbey Road* (1969), Elton John's *Goodbye Yellow Brick Road* (1973) and Oasis' *(What's the Story) Morning Glory?* (1995). Through a collection of interviews with the original music production, mastering and remastering practitioners directly involved with these iconic albums, we are able to gain comprehensive first-hand accounts of the creative and technical processes applied to produce the original musical

DOI: 10.4324/9781003177760-1

artefact and its subsequent remaster(s). Additionally, to research remastering practice as applied to live and orchestral recordings, a BBC broadcast of Mozart's *The Magic Flute* from 1962 is also examined. Furthermore, through comparative listening tests and digital audio analysis we are able to determine sonic differences between the original and remastered versions with the ability to directly connect these differences to the production and remastering processes followed. Finally, through these interviews we are also able to appraise and authenticate their thoughts regarding potential impacts of remastering practice on themes surrounding cultural heritage, legacy, authenticity and meaning.

Mastering and remastering overview

Mastering and remastering – what is it?

The traditional role of mastering was to correct and adjust the final mixed music recording, mainly focused on equalisation (EQ) and dynamics, as it was being transferred or cut from analogue tape to a record master that would then be used to stamp or press vinyl records for distribution. This was to ensure the vinyl records made commercially available to the public played back correctly on their home record players and sound systems both sonically as well as physically, for example, ensuring the stylus remained within the groove boundaries of the vinyl record. Although this process appears an important factor of music production, Temmer refers to it as insignificant in nature by comparison to the final musical product and merely a means to an end (1984). Alternatively, Nardi describes modern digital mastering practice as significant in determining musical production outcomes and a creative process, based primarily upon the ability of modern playback devices to better accommodate manipulation of EQ and dynamic range (2014).

Remastering is the practice of manipulating older recordings to make the sound optimal using modern playback systems and evolved through the introduction of the digital CD as a replacement for the analogue vinyl record format (Nardi, 2014). Initially, record companies simply transferred or stamped the original master tape used for vinyl reproduction on to the CD with minimal intervention, believing this was adequate (Sexton, 2009). However, as digital modern mastered recordings evolved alongside digital playback devices to produce a sonic quality consisting of a stronger emphasis on EQ and dynamics to optimise the format, early digital remasters of older recordings appeared to be lacking in these areas. As a result, remastering developed into the practice of using modern digital tools on older recordings to make them sound less lacking in these areas and sit more comfortably sonically beside the modern mastered ones. Remastering could therefore be described as the practice of creating a new digital replica or creative work from an existing analogue music artefact. When we consider the original musical artefact may be rich in musical history with an associated sense of cultural heritage, the amount and type of manipulation undertaken to create the digital replica becomes of crucial importance, particularly if the aim is to maintain the context, meaning and significance of the original work. There is also often the added pressure of the original work showing signs of degradation with a need to create a digital replica in order to preserve the original piece.

Remastering as an artistic and creative process

As defined previously, the traditional role of mastering was of a corrective and technical nature and primarily through the evolution of digital technology and tools available to the modern mastering engineer it has transformed into a more creative practice (Nardi, 2014; Rumsey, 2011). Elements concerning musical taste including sequencing, spacing, fading, equalisation and the choice to normalise an entire album to deliver slow ballads and faster more energetic rock songs at the same perceived loudness, could all be argued to be artistic decisions as opposed to technical ones (Shelvock, 2012). While the traditional aspects of the role may still exist, and there is an argument that the limited functions of mastering continue to be more correctional than creative, current research is demonstrating mastering as a creative dimension.

If we view the modern studio as more than simply a means to capture an artist's live sound, a creative environment with tools to digitally manipulate recordings, we place the mastering engineer at the 'intersection of what is widely acknowledged as the moment of creation, which is normally identified with songwriting, arranging and performance' (Nardi, 2014, p. 11). The concept of modern mastering as creative practice is also embodied in O'Malley's interview with producer and engineer Jeff Lynne, perhaps best known for his work with the Electric Light Orchestra (ELO) and production work on The Beatles' 1995 single release of 'Free as a Bird' and 'Real Love'. As he writes:

> Lynne often knew how he wanted the finished product to sound, and would drive every element of the process by personally overseeing the mastering sessions, and by adjusting EQ and compression to get the sound exactly as he wished. Lynne expressed a liking for the sonic quality that the inherent nature of vinyl imparted onto the sound of the master, sometimes feeling that a flat sounding mix was given a bit more punch by the process.
>
> *(O'Malley, 2015)*

Lynne's partiality for the process he describes above highlights the creative and artistic options available to the modern mastering engineer. The mastering engineer must therefore decide whether to 'transfer as much as possible of the sound of the digital master onto the vinyl disc or whether to enhance it in some way' and if the latter is decided this often results in creative discussion and consultation with others (Rumsey, 2014, p. 559). The remastering engineer operating within the digital environment appears to share the same artistic responsibilities as the modern mastering engineer, particularly with reference to working with vinyl, and often needs to apply these artistic and creative decisions to iconic analogue music recordings.

There is evidence to suggest that the creative and artistic choices available to the mastering and remastering engineer can influence musical attributes within a piece of recorded music through intervention. With French electro-house music for example, there is a tendency to 'maximize average amplitude and reduce dynamic range' during the mastering stage with an additional maximisation of lower end frequencies to ideally suit the large subwoofer speaker systems of the clubs and venues it will be played at (Shelvock, 2012, pp. 11–12). The ability to maximise average amplitude through dynamic signal processing onto the digital CD format has also been identified as a significant element of the grunge music genre of the late 1980s to early 1990s. Nirvana's 'Smells Like Teen Spirit' for example has been described as being 'distinguished by an increase in volume as much as anything else' (Vickers, 2010, p. 6). However, it is not just dynamic manipulation at the mastering and remastering stage leading to increases in volume, maximised average amplitude or reduced dynamic range of the remastered versions that can impact music genre. Digitally cutting and boosting certain frequencies at the mastering and remastering stage can also relate to genre, for example bright sounding for country and a reduced brightness for reggae (Shelvock, 2012). Additionally, it is likely that many musicians will object to tape hiss and other noise introduced through production long before they object to distortion added during the production process, which may be a desired feature in certain musical styles and genres and an artistic decision (Lagadec & Pelloni, 1983).

The creation of new digital formats has also led to further artistic and creative opportunities for mastering and remastering engineers, similar to the impact and introduction of the CD format to accommodate an increase in loudness unattainable on vinyl record production. Traktor Remix Sets for example is a digital format that allows DJs to mix, loop and process a stem master with isolated tracks which would not be possible with a stereo mix. This new format 'encourages new practices among mastering engineers, whose targets are not only music listeners but, specifically, music performers as well' (Nardi, 2014, p. 9). The role of the mastering and remastering engineer has therefore evolved with technology to accommodate the changes in musical tastes of the public and artists, and the manner in which recorded music is consumed. This evolution in mastering and remastering practice supports the view that modern mastering and remastering consists of two main parts: an aesthetic one and a technical one (Ojanen, 2015). However, the distinction and separation of technology and art in audio mastering appears impossible as both elements are intermittently entwined (Nardi, 2014).

Cultural and personal heritage in contemporary music

Within the fields of rock and contemporary music there appears to be an initial challenge of associating and classifying the cultural and heritage significance and impact using traditional definitions. These traditional definitions include the idea that the cultural artefact contains 'representations of custom, tradition and place that coalesce within the cultural memory of a particular national or regional context and fundamentally contribute to the shaping of the latter's collective identity' (Bennett, 2009, pp. 476–477). The mass-produced commercial properties of rock music appear to be in contrast to these traditional definitions; however, it can be argued that with regard to an ageing baby-boomer generation, rock and contemporary popular music is embedded within its cultural memory and generational identity (Bennett, 2009). There appears to be a need to digitally archive popular music recordings to protect local histories and cultural heritage (Baker, 2018), as well as provide new ways to interact with the art (Long & Wall, 2013). This digital archive practice therefore provides a link and continuity to the musical past (Cohen, 2016) in order for communities to have a close relationship to their music culture and heritage (Leonard, 2017).

This is evidenced by reference to the remastering engineers who worked on the 2009 Beatles' remastered CD release (see Chapter 2 for more details) as having 'been bestowed with the onerous task of sonically cleaning what is surely not just pop music but cultural heritage' (Clayton-Lea, 2009, p. 2). Another tangible account of the cultural heritage and significance of rock and popular music was the 'listed building' status afforded to the pedestrian crossing at Abbey Road (immortalised on the cover of The Beatles' album of the same name) by the official authorities (Roberts & Cohen, 2013).

Unlike in film restoration and remastering, where there is a clear attempt by industry organisations including the International Federation of Film Archives and the Association of Moving Image Archivists to develop a code of ethics to guide behaviour of practitioners to ensure any digital replica of a historical film artefact will enhance and not diminish our understanding of its original intended meaning and protect its perceived cultural heritage, music remastering practice appears devoid of an industry affiliated code of ethics (Mattock, 2010). This lack of a reference and benchmark for remastering engineers appears problematic, particularly when you consider the significance of their input into the creation of digital replicas and the constantly evolving array of digital tools and techniques available to them. Therefore, the capacity of the remastering engineer to falsify the original meaning and diminish cultural heritage of an original music artefact through its digital replication is arguably more likely in music as opposed to film, as there are no industry affiliated benchmarks or guidelines to reference and adhere to (Busche, 2006). To assist the music remastering engineer to deliver an acceptable digital replica, there is a belief that it would be beneficial to 'have information about what effects have been applied originally and how, in order to produce a remastered edition that sounds better, and not different from the original record' (Barchiesi & Reiss, 2010, p. 563). However, it is arguable that a technical blueprint regarding all of the nuances of the original production would be helpful considering there exists a pre-meditated approach amongst some practitioners to mainly focus on manipulation of the upper and lower ends of the useful spectral range only, when these are sometimes not the most needed adjustments (Brink, 1992).

With reference to the creation of a significantly different sonic and cultural digital representation of an original music artefact, it is important to define the practice of remixing, often confused with that of remastering (Bennett, 2009). Remixing, sampling and mashups all involve using digital technology to extrapolate sections and elements of an original music artefact to produce a new creative work. Therefore, remixes often bear little resemblance to the original musical artefact and purposely challenge Western conceptions of art and any sense of cultural heritage associated with the original iconic music recordings (Barham, 2014). The difference between the two practices and their respective impact on cultural heritage is portrayed in Bennett's interview below with record label Songworks' founder Mike (surname is not disclosed) using the Sistine Chapel painted ceiling as an analogy (2009). As he writes:

A.B.: Do you remix albums before putting them onto CD?

MIKE: We re-master, we don't remix … I think remixing is tampering with history. It's like nipping down to the Sistine Chapel and changing the colour scheme of it, y'know, as opposed to just cleaning it up.

A.B.: So you never do anything like that … even when the sound quality could be better by …

M.: Well that's re-mastering, not re-mixing. You're asking me about two different things. That's why [I use] the analogy with nipping down the Sistine Chapel. If you clean it up you can see it better that would be like re-mastering. But you don't change the colour scheme, 'cause that would be like re-mixing.

(Bennett, 2009, p. 485)

Remastering engineer Mike is adamant that remixing alters the meaning of the original work and interferes with the cultural heritage, identity and significance associated with the original work. Alternatively, Mike viewed remastering as primarily a restoration service used to enhance the cultural value and meaning of an original music artefact by making a digital replica that is more assessable for modern playback devices and with less noise and other sonic interference. However, choosing to remaster or remix an iconic analogue recording is not always a mutually exclusive exercise. Artist Jeff Lynne recently revisited his back catalogue and decided to re-record, remix, remaster and re-release various recordings to digitally improve performance and sonic elements (O'Malley, 2015). Lynne's decision to remix as opposed to remaster existing released songs from his back catalogue appears to challenge the artist's and listener's need or desire to maintain the original meaning and cultural heritage associated with the original recording. Furthermore, it could be argued that Lynne's choice to re-record and re-release various songs previously recorded and released, completely removing the original musical artefact from the production process, removes the perceived cultural heritage aligned with the original. The anticipated desire by listeners and artists to maintain a sense of cultural heritage and meaning associated with an original musical artefact through digital replication is therefore not conclusive.

Apart from the improvement of performance and sonic elements, another potential motivation for remastering would be to protect the original artefact from degradation, and hence guard against any loss of cultural heritage and meaning associated with the eventual demise of the original artefact (O'Malley, 2015). The original gates from Strawberry Fields, the Salvation Army children's home in Liverpool, UK, which inspired The Beatles' song 'Strawberry Fields Forever', have been replaced with replicas to produce an authentic experience without the risk of damage to the original artefact (Roberts & Cohen, 2013). The decision to replace these iconic gates to protect the original ones, to in a sense maintain a level of cultural heritage and meaning of the original artefact through a replication with no physical link with the original, tends to support the view that the community is satisfied culturally with this concept.

Another consideration in determining the cultural heritage and significance of an original musical artefact is the commercial push, often from record companies, to market and promote a digital replica or remastered release as somehow being superior in some way to the original release. In his study of The Beach Boys' iconic album *Pet Sounds* and their record company's marketing strategy for the remastered release, Bottomley suggests that their approach relied on the idea that 'the original was somehow flawed' and these flaws were rectified through 'remixing or remastering, reordered track lists and different artwork' (2016, p. 160). Furthermore, Bottomley states that the 'paratextual (cover art, liner notes, advertisements) materials suggested that the re-

release version is actually offered as being better—and more authentic—than any of its predecessors, including even the original' (2016, p. 160). If the digital replica is viewed as a superior product by comparison to the original music artefact, then cultural heritage, significance and meaning linked to the original artefact are also somehow flawed or diminished. However, as this was primarily a marketing campaign for commercial gain, the notion of the consumer perceiving digital replicas to be superior to the original is not clear. In an analysis of Swedish popular music sensation ABBA, it was stated that there was no conclusive evidence that the remastered product was superior to the original and in fact 'the irony is that the remastering of ABBA's catalogue has, with respect to dynamic range, probably gotten further from the master tapes with each new release' (Vickers, 2010, p. 7). Therefore, the ABBA remasters' format and sonic shape are also further away culturally with respect to the heritage, significance and meaning attributed to the original musical artefacts.

This attribution of meaning by the public to original music artefacts can be broken down further into what has been termed personal heritage. Barratt (2009) states that personal heritage was the creation of an archive of personal experience by ordinary individuals providing a deeper dimension to the historical event. As he writes:

> For example, the reality of the First World War suddenly made sense when viewed through the surviving letters, photographs, service record and associated unit war diary of a great grandfather who had left his home and family to fight in the trenches, rather than a depressing list of statistics, battles and political maps outlining the course of the war.
>
> *(Barratt, 2009, p. 10)*

If we were to place this example in musical terms, a relative delivering a detailed account of attending a Beatles' concert amidst the euphoria of Beatlemania, accompanied by a collection of photographs and ticket stubs potentially becomes more 'alive' and 'real' than perhaps witnessing a clip of the band performing the same concert on YouTube. Furthermore, personal heritage does not necessarily need to involve celebrities or well-established artists to be of use. According to Barratt, historians lament the lack of personal accounts of ordinary daily lives in medieval times to achieve a greater understanding of the conditions of life and encouraged 'people to keep track of their lives, so that their memories and stories are passed on to the next generation, and in many ways the internet offers the freedom, scope and technology to do so' (2009, p. 13). Therefore, digital technology offers musicians and artists, whether commercially successful or not, the opportunity to archive and distribute their experiences and recordings on various digital formats through digital media. Timothy concludes that 'people need the past to cope with the present, because patterns in the world make sense if we share a history with them' and the concept of personal heritage, particularly digitised personal accounts that can be distributed quickly and easily online, appears to allow that (1997, p. 752). It is therefore feasible to suggest that digital personal heritage can assist in the delivery and preservation of cultural heritage associated with established artists regarding time and place, particularly if a pre-digital era is examined. For example, if we explore the early career of The Beatles before they became famous it is likely there would be less imagery and recordings available as opposed to when they became successful, primarily due to the prohibitive (by today's standards) cost and cumbersome nature of photography and recording of the time. Therefore, through distributed digitised personal heritage collections of ordinary citizens who may have seen the band perform, played in a similar looking and sounding band, performed in comparable venues or lived in the same neighbourhood at the time, these materials arguably assist in creating a greater understanding of what life was like for the pre-famous Beatles.

Authenticity in contemporary music

The collection of alternate definitions and approaches regarding authenticity in contemporary music literature suggests a complexity of understanding and challenge of agreement amongst scholars. For Wu, Spieß and Lehmann, authenticity in music depended upon music genre, the era of performance and the social reference group, and comprised value indicators including 'credibility, fidelity, traditional and original, pure, real, serious, uniqueness, historically correct and being true to oneself' (2017, p. 443). Douglas claims that 'cultural constructions such as race, gender, and age all contribute to definitions of authenticity' (2016, p. 194). Authenticity therefore appears closely associated with time and place, particularly regarding era, race, age and social reference groups. Although these common themes concerning authenticity have been identified, Speers proposes that 'there is little agreement over what it is or, indeed, whether it even exists' (2017, p. 16). Furthermore, van Klyton concludes that although authenticity is often used to represent and determine the value of music, it is 'a conceptualisation of elusive, inadequately defined, other cultural, socially ordered genuineness' and, like most social values, subjective (2016, p. 107). Despite Speers' initial proposition of a degree of difficulty in understanding authenticity and agreement within scholarship, she defines three different approaches or ideologies concerning authenticity in music: the first where authenticity is inherent within the 'person, object, event or performance'; the second where authenticity is a 'socially agreed upon construct'; and the third where authenticity is 'produced through cultural activity and living them out' (2017, p. 16). Closely aligned to the ideology of authenticity as a social construct, Moore describes authenticity in contemporary music as being 'ascribed to rather than inscribed' that would suggest a causal connection between performer and audience (2002, p. 220). Furthermore, Moore states that it was perhaps more worthwhile to therefore identify 'who rather than what was being authenticated' (2002, p. 220). To identify who was being authenticated, Moore proposes that there are only three responses possible: 'the performer herself, the performer's audience, or an (absent) other who is being authenticated' (2002, p. 220). This ideology was developed further in Moore's study of authenticity as authentication, where he advises that authenticity in music can be broken down into three elements: authenticity in the first person, the second person and the third person (2002).

First-person authenticity 'arises when an originator (composer, performer) succeeds in conveying the impression that his/ her utterance is one of integrity, that it represents an attempt to communicate in an unmediated form with an audience' (Moore, 2002, p. 214). In essence the artist's performance conveys a sense of truth and honesty of their condition and circumstance and it is then up to the audience to either accept or reject this representation of authenticity. Wu, Spieß and Lehmann refer to this representation as 'personal authenticity' that 'addresses the relationship between musician and music' (2017, p. 446). Moore provides the example of artist Paul Weller's performance of 'The Changingman' whereby 'he employs gravelly vocals connoting a voice made raw from crying or shouting' and that 'the assumption here is that his listeners have personal experience of what gives rise to such crying and shouting and that, therefore, the result conjures up an active memory of the cause' (Moore, 2002, p. 212). Additionally, Moore suggests that Weller's use of 1960s vintage guitars and his practice of recording live with minimal overdubs, as opposed to current digital practice with more sophisticated options, further enhanced his representation of self as an authentic performer, someone who is exposing the truth and realness within (2002). This appears consistent with Davies who advises in his research concerning authenticity within orchestral music that 'the sound of an authentic performance will be the sound of those notes' which implies a pure performance or recording with historically accurate instruments (1987, p. 3). However, Connell suggests that authenticity through self-expression could be invented and constructed through behaviour and imagery that automatically implies continuity with some existing form or historic past (2002). Therefore, Weller's use of music instrumentation from the 1960s and its inherent imagery and practice of an older style and technique of music recording may be assigned a perception of authenticity from an audience not solely through Weller's expression of self but also from an 'automatic' memory of theirs from that time period. In Speers' study into authenticity concerning hip-hop music, she identifies a paradox between self and other (2017). She described the phenomenon that in order for rappers to gain acceptance and approval from others and potentially progress their respective careers, they conform to various concepts of authenticity work as predetermined by others and therefore they are not necessarily being true to themselves. As she writes:

> It raises the question of the extent to which a rapper's desire for authenticity is for him/herself, or for social validation among peers, or for acceptance from fans through perpetuating a particular view of what hip-hop is. The way in which rappers have to negotiate the tension between individual expression and community practices highlights the tension between 'rapper authenticity' and 'hip-hop authenticity'.
>
> *(Speers, 2017, p. 18)*

Second-person authenticity is described by Moore as being observed 'when a performance succeeds in conveying the impression to a listener that that listener's experience of life is being validated, that the music is "telling it like it is" for them' (2002, p. 220). This reference to authenticity appears consistent with Connell who advocates that authenticity is observed when the performer 'delivers a performance that serves to bring out fully its (inner) meaning and where listeners read this emotional meaning by bringing their personal experience to bear on the performance' (2002, p. 29). The audience or listener can therefore directly relate to the musical performance or recording as an authentic representation of their own life, or a perception of one. This makes this ideology of perceived authenticity a subjective matter according to Speers. As she writes:

> Therefore, the claim of authenticity made by or for a person, thing or performance has to be either accepted or rejected by relevant others. This is called the process of authentication. It calls attention to the importance of not just the intention of those wanting to be authentic, but how others receive and perceive them, which is a highly subjective affair.
>
> *(Speers, 2017, p. 16)*

In his study regarding the authenticity surrounding French folk music and dance, Revill concludes that over time and with the exclusion of access to high art definitions of culture, the folk 'generation by generation … transform folk music into something of true beauty created naturally and authentically from the unselfconscious actions of everyday folk' (2004, p. 205). Furthermore, Littlefield and Siudzinski contest that within communities representative of their 'own set of norms and values', songwriters gain particular respect in this community for their originality and authenticity (2011, p. 796). Popular music created naturally within a geographical location and/or community therefore assumes a sense of authenticity over time based upon participation by community members and the perception of other community members regarding what music is authentic to them. This sense of attaching a perception of local authenticity to music that is geographically based is, according to Holland, often exploited by 'media-driven stereotyping of musically prominent cities' to commercially construct a sense of place (2012, p. 126). For example, this is evident in the term Mersey Beat which was used to describe the music and sound emitting geographically from and commercially structured within the English city of Liverpool, predominantly started by The Beatles in the early to mid-1960s (Atkinson, 2011). However, van Klyton describes the potential diminished impact of cultural authenticity over time in his study concerning World Music (2016). He suggests that 'world music performances from West Africa, for example, could have a more modern (and less authentic feel) than it originally did 30 years ago' based upon the modernisation of some African territories and the perceived change in social, political and economic aspects (van Klyton, 2016, p. 108). Additionally, a societal and cultural shift towards modernisation is 'largely used to denote a dispersion and diffusion of values, a loss of aura and authenticity' (Jones, 2002, p. 213). If we revisit the example of Mersey Beat one could argue that the perception of authenticity linked towards original music produced today in Liverpool is perhaps less than that produced in the city 50 to 60 years ago based primarily upon globalisation, technological change and the city's transformation, as opposed to merely a change in musical trends.

Moore provides an example of second-person authenticity as the dedication by predominantly Celtic bands (U2, Simple Minds) in the 1980s against a backdrop of electronic synthesiser focused music to feature the guitar and adhere to 'traditional rock values for white urban bourgeois youth' to relate to as a better representation of themselves and identity (2002, p. 220). Wu, Spieß and Lehmann refer to this as 'cultural authenticity' and described it as concerned with the recipient's response, perception and reaction to the music in terms of being truthfully representative of a certain origin, culture, time and place (2017, p. 446). It could therefore be argued that perhaps this perception of cultural authenticity by the youth supporting these Celtic bands was also influenced by an opinion that technology-based music was an inauthentic alternative. As described by Connell, 'popular music that was placeless, created electronically and highly commercial, was regarded as a major disjuncture between producers and consumers, and a denial of authenticity' (2002, p. 38). In a similar example, the term Guitar-Pop was used to describe a subculture of the live music scene of Sydney in the early 1980s. It possessed a startling similarity to Mersey Beat in that it consisted of predominantly male bands with Beatlesque harmonies, haircuts and 1960s mod fashion, plying their trade with vintage looking and jangly and clean sounding Rickenbacker guitars through VOX AC-30 amplifiers (Easton, 2013). The use of historically accurate instruments was perhaps a purposeful strategy to attain a perception of cultural authenticity against the impending digital musical genre of 'New Wave' that was saturated in artificial MIDI instruments, sounds and automated drum machines. This appears consistent with Littlefield and Siudzinski's research that discovered that 'some brands represented a higher degree of authenticity and use of equipment from previous eras in music was one path to legitimacy and authenticity' (2011, p. 796). Therefore, it would appear that regardless of the perceived improvements with digital instruments and music production, the use of traditional and historic workflows and equipment suggest musicians justify their use of certain technologies as more authentic in relation to some real or imagined original.

Zagorski-Thomas argued that the inherent noise and playability associated with vintage and traditional instruments and recording equipment can produce a sound that could be perceived as authentic by an audience (2010). As he writes:

> Garage bands from the late 1950s onwards have produced rough and unpolished recordings and this has led to it being embraced as a production aesthetic in itself. If the dilettante approach is deliberate, however, then it takes on additional meaning: a professional quality recording may become a signifier for the 'establishment' and the rejection of it through choosing a lo-fi approach becomes a political statement of difference.
>
> *(Zagorski-Thomas, 2010, p. 262)*

Inherent noise associated with the original recordings due to technological limitations of the time, as opposed to being part of the original music composition and expression, has therefore become embedded in the ways that musicians and listeners construct some recorded music as more authentic than others. The rejection of establishment to produce a 'lo-fi' recording as described above would potentially have been an economic point of difference as well as political. With respect to remastering, where one of the prime objectives is to remove and clean up inherent noise associated with the original recording to make it sound more 'modern' and able to sit comfortably next to a modern recording, it therefore becomes a question of whether removing noise is in fact removing key sonic signifiers that listeners associate with concepts of authenticity in recorded rock music (Bennett, 2009). The correlation between inherent analogue noise and authenticity is perhaps best evidenced in Zagorski-Thomas's research that found that current professional DJs using modern digital music production tools deliberately introduced noise associated with vintage playback and recording devices (stylus crackle from a record player for example) on modern recordings (2010). We therefore have a state where noise associated with vintage recording and playback devices is purposely being reintroduced deliberately to new music recordings to instil a sense of cultural authenticity for the listener. Again, this deliberate digital manipulation to recreate analogue noise appears to be a strategy to perhaps provide security for the listener against their perception of 'artificial' modern music production and performance.

Moore describes third-person authenticity as 'when a performer succeeds in conveying the impression of accurately representing the ideas of another, embedded within a tradition of performance' (2002, p. 218). The perception of authenticity therefore appears to imply the ability for one to mirror the original intent of the composer and/or the recording of a composition or live performance from an established collection of acceptable performances. This seems consistent with Davies' study into authenticity in classical music performance where he states that 'authentic is used to acknowledge the creative role of the performer in faithfully realising the composer's specifications' when the musical performance closely matched and realised the composer's original score and the performance could be judged against a set of other accurate performances (1987, p. 1). Moore claims that third-person authenticity was closely linked to first-person authenticity in that it required the audience to perceive the artistic performance of the musician as an authentic depiction of self, as well as an expression that captured the style of another performer(s) that was also perceived as authentic (2002). Moore provided the example of Eric Clapton performing Robert Johnson's blues compositions as third-person authenticity. As he writes:

> In performing Johnson's 'Crossroads' with Cream, not only do we interpret Clapton conveying to his audience that 'this is what it's like to be me' but, doubly vicariously, that 'this is what it was like to be Johnson', with all the pain that implies: '[The blues] comes from an emotional poverty … I didn't feel I had any identity, and the first time I heard blues music it was like a crying of the soul to me. I immediately identified with it'.
>
> *(Clapton quoted in Coleman 1994, p. 31, cited in Moore, 2002, p. 215)*

Moore's work on authenticity in music suggests that as it is ascribed there should be a greater focus on who is being authenticated as opposed to what. His first-, second- and third-person authenticity ideology represent how a perception of authenticity can vary between the performer and the audience across common themes of time, place, gender, ethnicity and technological mediation. It is the social construct of authenticity within this ideology that, according to Speers, suggests authenticity as evolutionary as the 'continual quest for a creative voice has the effect of destabilizing the image of the authentic' (2017, p. 17). This sense of authenticity as evolutionary seems important with respect to remastering when you consider the constant replication and manipulation of a musical artefact and the potential for undermining the image of the original master tape. The opposing view is that authenticity in music is inherent within the performance, artist or object, and this would suggest a greater prominence concerning what as opposed to who is being authenticated, resulting in a more static sense of authenticity. This static approach to authenticity has been described as 'type authenticity' and depicts 'how well the artist follows and conforms to the features and elements of an existing genre. In this respect, deviating from a genre indicates a de facto decrease in type authenticity' (Mattsson, Peltoniemi & Parvinen, 2010, p. 1358). Examples of type authenticity artists would include The Monkees, Spice Girls and those who emerge from reality TV shows like Australian Idol where they are manufactured and designed accordingly within a specific genre ready for marketing and distribution to a predetermined audience (Mattsson et al., 2010).

Commercial considerations and the 'loudness wars'

In reference to music and cultural heritage it has been argued that 'music heritage increasingly encompasses a range of practices that are not reducible to the music itself but linked to a wider social, cultural and economic processes surrounding the production and consumption of popular music' (Roberts & Cohen, 2013, p. 2). It is these economic processes, often beginning with a shared concept and goal within a record company regarding the production and consumption of music, that lead in part to the decision to digitally remaster and re-release older analogue recordings to the public (Rumsey, 2012). Therefore, while there may exist debate about the artistic and cultural significance of remastering older analogue recordings, 'many such recordings do have a commercial value to those who restore them' (Brink, 1992, p. 7). For example, Apple Corps' manager Neil Aspinall, when discussing the company's decision to remaster and re-release The Beatles' entire back catalogue on CD in 2009, referred to not only the perceived sonic improvements regarding the release but also described them as packages and the need to 'get proper booklets to go with each of the packages' (Benzine, 2006, p. 1). The package is therefore representative of far more than just the music itself. While there may be a tendency for commercial organisations to focus purely on commercial outcomes from revenue gained by releasing remastered versions of their artists' catalogue of recordings, the 'latest wave of Jimi Hendrix reissues marks the fourth time the catalogue has been remastered for CD', there is also an alternate view (Richardson, 1997, p. 96). According to Clayton-Lea, there was a belief from the remastering engineers involved in The Beatles' 2009 CD remaster project that it was not purely a commercial exercise (2009). As he writes:

> In terms of The Beatles' 1960s masters – they were never re-mastered, simple as that. What you got in the 1960s was what you got in the 1980s, and to bring them up a little bit, sonically, we re-mastered them. So it's not a money issue at all. Having said that, it's not really for me to say because we just do the job – we're told to do it. But my personal opinion is that it isn't about the money. Absolutely not.

It is important to note that although the arrival of the CD initially was seen as commercially prosperous for record companies with respect to selling their back catalogue in the new format, there were also critics of this decision at the time. Maurice Oberstein from Polygram UK voiced his concerns at an industry conference that the record companies were literally giving away the master tapes and that the ensuing recordable CD format would lead to a black-market economy of illegal copying and distribution (Sandall, 2007). The new CD format also allowed it to contain recordings at a greater loudness level in playback as compared to releases on analogue vinyl and cassette tape (Nielsen & Lund, 1999). It is believed the initial demand to create louder recordings on CD was derived from the commercial interests of FM radio broadcasters in the early

1990s as they sought a competitive advantage to broadcast at a higher level than their opponents (Vickers, 2010). This mastering and remastering phenomenon is known as 'hyper compression' and resulted in the well-documented and researched 'loudness wars' which suggest these louder produced recordings come at a cost of broadcast quality and reduced dynamic range (Hjortkjaer & Walther-Hansen, 2014; Nielsen & Lund, 1999; Vickers, 2010; Walsh et al., 2011).

> Pop/rock albums released towards the middle and end of the 1990s (and onward) are severely compressed and comparatively louder than records of the past, averaging 4dBFS less in dynamic range than noticeably compressed masters by the likes of the Beatles, Motown, and others.
>
> *(Shelvock, 2012, p. 55)*

To make remasters commercially viable and appealing to the consumer, record companies often use a marketing strategy designed to 'sell the notion to existing fans that the reissue is in some way superior to any previous edition' (O'Malley, 2015, p. 3). In Vickers' study of the commercial decisions behind Universal Music Group's approach to marketing their ABBA back catalogue in 2001 and again in 2005, it was revealed that 'as long as Universal Music Group has financial incentive to keep revisiting the Abba [*sic*] catalogue, they need to appeal to the notion that each time they are getting closer and closer to the original, bringing out further unheard details' (2010, p. 7). There also exists a commercial consideration and 'belief that louder songs sell better' that ensure increased loudness levels on the remastered releases as compared to the original recordings often with accompanying packaging detailing justification for the increase (Vickers, 2010, p. 5). However, there is evidence to suggest criticism that hyper compression and the resulting extra loudness has damaged the audio quality of remasters by decreasing dynamics and thereby reducing the emotional elements of the song as well as leading to listening fatigue (Vickers, 2010).

Do listeners notice differences between original and remastered releases?

Although the 'loudness wars' have been well documented – and increased levels of loudness associated with digital formats have been well defined – it is still unclear whether the perceived sonic differences between an original and remastered release are noticeable by the consumer. In his comparative study of various remastered releases of The Beatles' song 'All My Loving', Barry questions whether the average consumer would notice the perceived sonic differences he identified through his audio analysis (2013). In another research project to determine whether average listeners experience the claimed negative effects of compression in remasters, it was determined that the result was unclear and in fact 'listeners are less sensitive to even high levels of compression than commonly claimed' (Hjortkjaer & Walther-Hansen, 2014, p. 39). Although not directly related to remastering, there is evidence to suggest listeners on occasion tend to exaggerate their aural perceptions. Record reviews from the early 1920s included phrases such as 'totally natural', 'indistinguishable from the live performance' and 'never to be surpassed quality', which would appear to contrast the inherent noise associated with 1920s recordings and playback devices (Temmer, 1984,

p. 2). Furthermore, there is verification of how easy it is for mastering engineers to persuade people something has changed when it has not and that experienced listeners can be confused over compression through incremental changes in perceived loudness (Rumsey, 2010; Vickers, 2010).

One consideration for the listeners' apparent lack of ability to adequately identify differences between remastered and original versions is the disparity between the high-quality studio environment where the remaster is produced, and lower quality consumer playback devices. According to The Beatles' remastering engineer Rouse, there is a consistent generational approach to try to provide the best sound quality possible at Abbey Road Studios and he lamented the inferior quality of modern playback devices as the listener only hears part of what is there (cited in Sexton, 2009). A related consideration is the impact that mono and stereo recording formats have on the remastering process and what listeners are able to access. The Beatles band members have been reported as saying that *Sgt. Pepper's Lonely Hearts Club Band* – revered by many as a masterpiece of technical innovation and construction for the time – was best listened to in mono, although only recently has the digital mono version been made available (Kot, 2009). In contrast, Paton and McIntyre undertook a study comparing average consumers listening to mastered and unmastered pieces of music and concluded listeners preferred the mastered works (2009). Additionally, when fans and consumers of Metallica's *Death Magnetic* album were confronted with a choice of either a Game Hero or a much louder and more compressed CD version, 'over 21,000 people signed an online petition asking Metallica to remaster the CD with less compression' (Vickers, 2010, pp. 6–7). This desire for less compression was consistent with the view that louder does not mean better and the heavily compressed releases of Hendrix have been described by listeners as making 'the already trebly guitar of House Burning Down too bright, and it heightens the background buzz and hiss on both Voodoo Child and Voodoo Chile' (Richardson, 1997, p. 96).

The above research and examples would suggest that listeners can tell the sonic differences between mastered and unmastered versions, in particular with respect to heavily compressed as compared to less compressed. It is therefore conceivable that this capacity of the listener to determine difference has been attributed to the current trend in increased LP sales as the listener perceives vinyl has a warmer, more nuanced sound as compared to CDs and digital downloads 'perhaps because of the necessity of using finesse to work around vinyl's physical limitations regarding signal levels' (Vickers, 2010, p. 14). However, Uwins 'rejects the hypothesis that audio quality is the sole defining factor' in his study comparing vinyl and CD recordings and attributed the resurgence in LP sales more to an 'individual's appreciation of other attributes of vinyl such as the artwork, sleeve notes, or even their past experiences' (2015, p. 5). Overall, the research in this area is somewhat contradictory or complex, with some research suggesting that listeners cannot easily identify sonic differences between remastered and original releases (Barry, 2013; Hjortkjaer & Walther-Hansen, 2014), and other research suggesting that they can (Paton & McIntyre, 2009; Vickers, 2010).

Digital audio analysis methodology

To explore remastering practice further, I chose to generate and analyse quantitative data in the form of digital audio music files. This allowed me to not only compare the remastered and original music recordings against each other to identify any differences and consistencies, but also to examine if there was any correlation between the perceived sonic qualities provided qualitatively by the case study participants and the statistical impression of the quantitative data.

The method I adopted closely followed the data analysis techniques I used in my previous research (Bruel, 2019; Bruel, 2021a; Bruel, 2021b), as well as techniques used by Barry (2013) to compare The Beatles' re-releases, and the analysis of recordings undertaken by O'Malley (2015). The methods used are outlined below with greater detail of the overall process described in the relevant case study sections. The first step in the process involved creating digital WAV file versions of the vinyl recordings of The Beatles' *Abbey Road* and Elton John's *Goodbye Yellow Brick Road*. It was important to find and use an authentic first-generation copy as this would have most likely emanated from the same original master tape used to create the 1995 and 2013 remastered CDs (Rick O'Neil, personal communication, 9 November 2015). I managed to track down first pressing versions through UK-online record dealer Better on Vinyl as verified by owner Chris Edgecombe. Chris has been collecting records for a long time and was able to verify both the *Abbey Road* and *Goodbye Yellow Brick Road* albums based upon key characteristics. As he writes:

> The *Abbey Road* album can be identified as a first pressing by the fully laminated sleeve without flaps on the back side. The front cover has the drain showing, the rear cover has the Apple logo aligned with the text, and the track 'Her Majesty' is not listed on the rear cover or the label on the record. Furthermore, the matrix area (the non-grooved area between the last track on the record and the label) on Side A is encoded with YEX 749. Records were often rushed out to get ahead of the pack (if the release coincided with another big band like The Rolling Stones or Led Zeppelin or Fleetwood Mac etc there was a degree of carelessness which makes it all the more interesting).
>
> The *Goodbye Yellow Brick Road* album can be identified as a first pressing by the tri-fold out sleeve, it was made in England, the black labels with yellow logo and silver text. The matrix area on side 1 has DJLPD 1001 A-3 encoded on it, which is within the range of codes used for versions released in 1973. The vinyl also has a red translucent colour which can only be seen when holding it up to a bright light. I believe this had something to do with a petrol shortage at the time, but I can't find any information on it. Most Elton John albums released around the early 1970s have the same translucent red vinyl.
>
> *(Chris Edgecombe, personal communication, 18 March 2022)*

To create a digital copy of these albums, I first played them on a Rega Planar 1 Plus turntable with a built-in phono stage high specification pre-amplifier. From there I sent the signal into Pro Tools with a bit depth and sample rate of 24bit 48kHz using industry standard studio equipment. I added around 6dB during this stage to bring the signal more in line with the CD remasters for a fairer comparison. These recordings were then 'bounced' out of Pro Tools as WAV files with a bit depth and sample rate of 16bit 44.1kHz, the CD standard, to enable direct comparison with the CD remasters, resulting in the following digital files for analysis (see Table 0.1).

TABLE 0.1 Digitised and remastered file types

Artist	*Album*	*Transferred format*	*Remastered format*	*Remastered format*
The Beatles	*Abbey Road*	1969 original first vinyl pressing as 16bit/44.1kHz bit depth and sample rate WAV file(s)	1986 CD remastered release as 16bit/44.1kHz bit depth and sample rate WAV file(s)	2009 CD remastered release as 16bit/44.1kHz bit depth and sample rate WAV file(s)
Elton John	*Goodbye Yellow Brick Road*	1973 original first vinyl pressing as 16bit/44.1kHz bit depth and sample rate WAV file(s)	1995 CD remastered release as 16bit/44.1kHz bit depth and sample rate WAV file(s)	2014 CD remastered as 16bit/ 44.1kHz bit depth and sample rate WAV file(s)
Oasis	*(What's the Story) Morning Glory?*	1995 CD release as 16bit/44.1kHz bit depth and sample rate WAV file(s)	2014 CD remastered as 16bit/ 44.1kHz bit depth and sample rate WAV file(s)	
Mozart	*The Magic Flute*	1962 original tape recording of broadcast as 16bit/ 44.1kHz bit depth and sample rate WAV file(s)	2015 remastered CD as 16bit/ 44.1kHz bit depth and sample rate WAV file(s)	

The next step was to analyse the data for each artist's set of recordings. For example, the digitised copy of the original 1973 released Elton John *Goodbye Yellow Brick Road* album could be analysed against its subsequent 1995 and 2014 CD remasters as they were both now in the same digital WAV file format. Similar to Barry's study, the digital analysis focused on a comparison of loudness signal and shape through the graphical representation of the visual waveforms within the Pro Tools environment (2013).

From there, true peak (all peaks measured regardless of the duration of the peak), root means squared (RMS) (the standard average value in dBFS measured over the entire recording and is used primarily to display the 'average' level of loudness overall), loudness units relative to full scale (LUFS) (unit measurement of average loudness which takes into consideration human perception) and dynamic range was measured through inputting each digital audio file into the MATT DROffline Mk11 meter. All true peak, RMS and LUFS measurements attained correspond to the International Telecommunications Union (ITU) Broadcasting Services (BS) 1770 global specification recommendation and the dynamic range score is generated through calculating deviations of loudness distribution within a complete song as well as differences between average and peak loudness (MAAT Inc., 2021). The dynamic range algorithm measures dynamic density in music recordings (MAAT Inc., 2021).

Similar to the analysis undertaken by O'Malley (2015), I used the Voxengo SPAN Plus Fast Fourier Transform (FFT) audio spectrum analyser plug-in within the Pro Tools environment to attain a visual representation of signal level across the frequency spectrum. I used the Real Time Average (RT AVG) measurement type to produce averaged spectrum representations. I set up each consolidated WAV file on a separate track within Pro Tools and inserted the Voxengo SPAN Plus plug-in on each channel. I then created an auxiliary track with the Voxengo SPAN Plus plug-in and routed the tracks through the Voxengo SPAN Plus routing network so that I could measure visually the separate versions against each other in the one image. It is important to note that these images only provide a brief snapshot in time on the various frequency/volume levels across all of the releases although it is still useful as a frequency plot for comparison.

This audio analysis method adopted was significant and useful as it provided statistical and visual technical audio representations of the differences and consistencies between original and remastered recordings. It also provided statistical evidence and data to either support or question the sonic differences between the original and remastered version as perceived and described by the case study participants. The resulting comparisons are discussed in detail in the respective case study section for each artist recording:

- Midlife Crisis Records: www.midlifecrisisrecords.com
- My staff profile at University of Lincoln: https://staff.lincoln.ac.uk/fdfed3c0-0478-4209-9e6f-dfc2d4b68905
- LinkedIn: www.linkedin.com/in/stephenbruel/

Bibliography

Atkinson, P. (2011). The Beatles on BBC Radio in 1963: The 'Scouse' inflection and a politics of sound in the rise of the Mersey Beat. *Popular Music and Society*, 34(2), 163–175. doi:10.1080/03007760903268809.

Baker, S. (2018). *Community Custodians of Popular Music's Past: A DIY Approach to Heritage.* Routledge.

Barchiesi, D., & Reiss, J. (2010). Reverse engineering of a mix. *Journal of the Audio Engineering Society*, 58(7/8), 563–576.

Barham, J. (2014). 'Not necessarily Mahler': Remix, samples and borrowing in the age of wiki. *Contemporary Music Review*, 33(2), 128–147.

Barratt, N. (2009). From memory to digital record: Personal heritage and archive use in the twenty-first century. *Records Management Journal*, 19(1), 8–15. doi:10.1108/09565690910937209.

Barry, B. (2013). *(Re)releasing the Beatles.* Paper presented at the Audio Engineering Society Convention 135, New York, USA.

Bennett, A. (2009). 'Heritage rock': Rock music, representation and heritage discourse. *Poetics*, 37(5–6), 474–489.

Benzine, A. (2006, 15 April). Court hears Apple Corps plan for Beatles digital release. *Music Week*, 4.

Bottomley, A. J. (2016). Play it again: Rock music reissues and the production of the past for the present. *Popular Music and Society*, 39(2), 151–174.

Brink, R. S. J. (1992). *Empirical Methods in Restorative Processing of Historical Recordings.* Paper presented at the Audio Engineering Society Convention 92, Vienna, Austria.

Bruel, S. (2021a) Remastering Sunnyboys. In J. P. Braddock, R. Hepworth-Sawyer, J. Hodgson, M. Shelvock & R. Toulson (Eds.), *Mastering in Music* (pp. 155–173). Routledge.

Bruel, S. (2021b) Remastering the independent past. In V. Sarafian (Ed.), *The Road to Independence. The Independent Record Industry in Transition* (pp. 51–97). University of Toulouse 1 Capitol.
Bruel, S. (2019) Nostalgia, authenticity and the culture and practice of remastering music. (Doctoral dissertation). Retrieved from https://eprints.qut.edu.au/129568/.
Busche, A. (2006). Just another form of ideology? Ethical and methodological principles in film restoration. *The Moving Image*, 6(2), 1–29.
Cohen, S. (2016). *Decline, Renewal and the City in Popular Music Culture: Beyond the Beatles.* Routledge.
Clayton-Lea, T. (2009, 25 August). The magical remastering tour. *Irish Times*, p. 12.
Connell, J. (2002). *Sound Tracks: Popular Music, Identity, and Place.* Routledge.
Davies, S. (1987). Authenticity in musical performance. *British Journal of Aesthetics*, 27(1), 39–50.
Deruty, E., & Tardieu, D. (2014). About dynamic processing in mainstream music. *Journal of the Audio Engineering Society*, 62(1/2), 42–55.
Douglas, S. (2016). *Voicing Girlhood in Popular Music: Performance, Authority, Authenticity.* Routledge.
Easton, M. (2013). Down the lane: Sydney's DIY music scene. *The Lifted Brow* (17), 42.
Hjortkjaer, J., & Walther-Hansen, M. (2014). Perceptual effects of dynamic range compression in popular music recordings. *Journal of the Audio Engineering Society*, 62(1/2), 37–41.
Holland, M. (2012). An 'aesthetic of sorts': Technological advancement, authenticity and music production practices in Dunedin, New Zealand. In *Routes, Roots and Routines: Selected papers from the 2011 Australia/New Zealand IASPM Conference.* International Association for the Study of Popular Music Australia New Zealand Branch.
Jones, S. (2002). Music that moves: Popular music, distribution and network technologies. *Cultural Studies*, 16(2), 213–232. doi:10.1080/09502380110107562.
Katz, R. A. (2007). *Mastering Audio: The Art and the Science.* Elsevier/Focal Press.
Kot, G. (2009, 6 September). Is it worth the price? *Chicago Tribune*, p. 4.
Lagadec, R., & Pelloni, D. (1983). *Signal Enhancement via Digital Signal Processing.* Paper presented at the Audio Engineering Society Convention 74, New York, USA.
Leonard, M. (2017). *Gender in the Music Industry: Rock, Discourse and Girl Power.* London: Routledge.
Littlefield, J., & Siudzinski, R. (2011). Is that a real song or did you just make it up? Styles of authenticity in the cultural (re)production of music. *Advances in Consumer Research*, 38, 1.
Long, P., & Wall, T. (2013). *Media Studies: Texts, Production, Context*, 2nd edn. Routledge.
MAAT Inc. (2021). *DROffline MkII: User Manual.* Retrieved from https://www.maat.digital/support/.
Mattock, L. K. (2010). From film restoration to digital emulation: The archival code of ethics in the age of digital reproduction. *Journal of Information Ethics*, 19(1), 74–85.
Mattsson, J. T., Peltoniemi, M., & Parvinen, P. M. T. (2010). Genre-deviating artist entry: The role of authenticity and fuzziness. *Management Decision*, 48(9), 1355–1364. doi:10.1108/00251741011082107.
Moore, A. (2017). *Rock: The Primary Text – Developing a Musicology of Rock.* Routledge. doi:10.4324/9781315209791.
Moore, A. (2002). Authenticity as authentication. *Popular Music*, 21(2), 209–223.
Morey, J. (2009). Arctic Monkeys – The demos vs. the album. *Journal on the Art of Record Production* (04).
Nardi, C. (2014). Gateway of sound: Reassessing the role of audio mastering in the art of record production. *Dancecult*, 6(1), 8–25.
Nielsen, S. H., & Lund, T. (1999). *Level Control in Digital Mastering.* Paper presented at the Audio Engineering Society Convention 107, New York, USA.
O'Malley, M. (2015). The definitive edition (digitally remastered). *Journal on the Art of Record Production* (10).

Ojanen, M. (2015). Mastering Kurenniemi's rules (2012): The role of the audio engineer in the mastering process. *Journal on the Art of Record Production* (01).
Paton, B., & McIntyre, P. (2009). *Audio Mastering: Experimenting on the Creative System of Music Production*. Paper presented at The Second International Conference on Music Communication Science, Sydney, Australia.
Revill, G. (2004). Performing French folk music: Dance, authenticity and nonrepresentational theory. *Cultural Geographies*, 11(2), 199–209. doi:10.1191/14744744004eu302xx.
Richardson, K. (1997). Still remastering, still dreaming. *Stereo Review*, 62(9), 96.
Roberts, L., & Cohen, S. (2013). Unauthorising popular music heritage: Outline of a critical framework. *International Journal of Heritage Studies*, 20(3), 241–261.
Rumsey, F. (2014). The vinyl frontier. *Journal of the Audio Engineering Society*, 62(7/8), 559–562.
Rumsey, F. (2012). Pound of cure or ounce of prevention? Archiving and preservation in action. *Journal of the Audio Engineering Society*, 60(1/2), 79–82.
Rumsey, F. (2011). Mastering: Art, perception, technologies. *Journal of the Audio Engineering Society*, 59(6), 436–440.
Rumsey, F. (2010). Mastering in an ever-expanding universe. *Journal of the Audio Engineering Society*, 58(1/2), 65–71.
Sandall, R. (2007, August). Off the record. *Prospect Magazine*.
Sexton, P. (2009, 12 September). Repaving 'Abbey Road'. *Billboard*, 121, 24.
Shelvock, M. (2012). Audio mastering as musical practice. (Master of Arts), The University of Western Ontario.
Speers, L. (2017). *Hip-hop Authenticity and the London Scene Living Out Authenticity in Popular Music*. London: Routledge.
Temmer, S. F. (1984). *The Enjoyment of Recorded Music in a Technologically Oriented Society*. Paper presented at the Audio Engineering Society Convention 1r Australian Regional Conference, Melbourne, Australia.
Timothy, D. J. (1997). Tourism and the personal heritage experience. *Annals of Tourism Research*, 24(3), 751–754. doi:10.1016/S0160-7383(97)00006-6.
Uwins, M. (2015, 7–10 May). *Analogue Hearts, Digital Minds? An Investigation into Perceptions of the Audio Quality of Vinyl*. Paper presented at the Audio Engineering Society Convention 138, Warsaw, Poland.
van Klyton, A. (2016). All the way from authenticity and distance in world music production. *Cultural Studies*, 30(1), 106–128. doi:10.1080/09502386.2014.974642.
Vickers, E. (2010). *The Loudness War: Background, Speculation, and Recommendations*. Paper presented at the Audio Engineering Society Convention 129, San Francisco, California, USA.
Walsh, M., Stein, E., & Jot, J.-M. (2011). *Adaptive Dynamics Enhancement*. Paper presented at the Audio Engineering Society Convention 130, London, UK.
Wu, L., Spieß, M., & Lehmann, M. (2017). The effect of authenticity in music on the subjective theories and aesthetical evaluation of listeners: A randomized experiment. *Musicae Scientiae*, 21(4), 442–464. doi:10.1177/1029864916676301.
Zagorski-Thomas, S. (2010). The stadium in your bedroom: Functional staging, authenticity and the audience-led aesthetic in record production. *Popular Music*, 29(2), 251–266.

1

REMASTERING FROM VINTAGE FORMATS

Introduction

To provide a comprehensive account of remastering practice and to ascertain how this practice varies with respect to the configuration of the original musical artefact, it is important to explore not only popular commercial media formats and masters (generally tape or digital), but also examine other older and less-used formats (in some cases non-commercial). In order to do this, I interviewed Will Prentice from the British Library sound archive and renowned recording and remastering engineer Mike Dutton (The Cure, Michael Nyman) to describe remastering practice, audio restoration and digitisation processes as applied to wax cylinders, acoustically and electrically recorded shellac discs, acetates/lacquers, vinyl, commercial and non-commercial tape (reel-to-reel and cassette), and other formats (Dictabelt, Tefifon and Quad eight-track cartridges).

Will Prentice (background)

I grew up on a farm in Scotland and naturally transcended towards playing the bagpipes. From there I played in a number of pipe bands, punk bands and also gained experience in music production in various studios. I completed a classical music degree which led me to take on and finish an ethnomusicology-based master's degree. While studying my master's, I volunteered at the British Library sound archive to help catalogue their world and traditional music section. Funding allowed me to then work on digitising ethnographic wax cylinders for six months, which led to a permanent job as an audio engineer with the British Library sound archive for the past 21 years.

I am currently the Training and Dissemination Manager for the Library's Unlocking Our Sound Heritage project where I am responsible for training staff to participate in the project. As a team we focus on caring for and preserving audio artefacts across various formats spanning back to the 1880s, digitising assets in an attempt to make them available for researchers and to minimise further degradation of these precious materials.

DOI: 10.4324/9781003177760-2

Mike Dutton (background)

I'm originally from Harrow Weald, Greater London. During my youth I collected 78s, LPs, old wireless radios, tape recorders and record players. In 1976, at the age of 16, I joined Morgan Studios in Willesden, northwest London as an apprentice where I was trained to become a recording engineer. I worked with many artists, including The Cure, and in 1981, I left to work in theatre sound in the West End on productions by Andrew Lloyd Webber and Cameron Mackintosh. In 1985, I joined Pye/PRT Studios as a recording engineer, where I also ran the CD editing and mastering department. While in this role, I collaborated with CEDAR on the development of their new audio restoration system, which I then used in my own remastering work. In 1990, I left Pye/PRT and joined EMI Classics, based at Abbey Road Studios, London where I worked as a remastering engineer on CD reissues of vintage 78s and tape recordings. I also continued work as a recording engineer with a number of artists including Michael Nyman on his film score for Jane Campion's film *The Piano*.

In 1993 I set up my own record company, Dutton Laboratories, which specialised in reissuing vintage classical recordings on CD, which I remastered from 78s using CEDAR's Cambridge system. During this time, I also worked as a recording engineer for numerous classical labels, recording orchestras, ensembles and soloists often in historic venues. In 1999, I set up the Dutton Epoch record label to focus on releasing previously unrecorded works by English and American composers. Since 2008, I have remastered and reissued pop, jazz, rock and middle-of-the-road albums licensed from various major labels' catalogues through my Vocalion label. In 2015, I began reissuing recordings in the SACD stereo/multi-channel hybrid format, and more recently I have been remastering back-catalogue material using the Dolby Atmos and Sony 360 Reality Audio surround-sound systems.

Wax cylinders (Will Prentice)

Time periods of recordings

Solid wax cylinders first appeared in the summer of 1888, and the earliest ones we have in the sound archive are from the mid to late 1890s. The majority of our collection predates the First World War although some of them are from the 1920s and includes both commercially released (popular music of the time) and non-commercial cylinders (used for research and linguistic pursuits). It is interesting to note that although commercial discs replaced cylinders in the home entertainment market in the UK in 1908, as a technology, cylinders were still really useful for linguistic and/or music researchers and continued as a research recording format long after other people stopped using them. Disc recording wasn't really available for non-commercial non-professional purposes until the lacquer acetate disc appears in 1934.

The sound archive tends to prioritise the digitisation process for our non-commercial over our commercial stock. The main reason for this strategy is that the content is often unique (they may be a one-off or form part of a limited run) and if we were unable to save it, it would be lost forever. For example, we have ethnographic music and linguistic cylinder recordings of cultures from all around the world, including rare ones and those that may be extinct, so for researchers in particular, these materials can be very valuable. Additionally, these cylinders were often made of brown wax, which is softer, degrades quickly and attracts more mould than their commercial counterparts. Furthermore, there are often numerous copies available of the commercially released cylinders both within our archive and in external organisations.

Identifying and cleaning the cylinders

The first thing we do is open the cylinder box, consisting of a cardboard lid and tube, and try to determine if the lid and tube belong together. Often the lid, which is usually the only place you will find any writing, is connected to the wrong tube and or is misplaced. This makes identification challenging and so you don't know what is on the cylinder until it is played. We are basically looking for any contextual information that will help us match the correct cylinder, tube and lid, and this often requires further research, so as to assist us with attaining an accurate transfer of the recording described.

It is important to clean the cylinders to optimise playback and to obtain the best digital transfer we can, so the next stage involves carefully removing the cylinder from the tube with two fingers and examining its condition. We are primarily looking for cracks and evidence of mould. Given that these cylinders have often been stored in humid places, around the world, potentially for 120 years, mould is unfortunately common. We use protective clothing and brush away loose dry mould (if it is slimy mould it needs to be dried first) and dirt using an extractor vacuum to remove the debris. We then play it through on our playback system, and have found that as the groove is vertical, the stylus removes a lot of the mould so that the second playthrough sounds vastly superior. We tend to avoid using mould removal chemicals as they can often reveal the damage beneath the mould which can make playback worse.

Playing the wax cylinder

The wax cylinder player we use is custom built and largely designed by us. Our engineers Peter Copeland and Ted Kendall created the design, and it was built by a guy called Peter Posthumous. It is based around a mandrel type spinning cone, that is driven by a

FIGURE 1.1 Brown wax Edison 'Concert' cylinder being fitted to the mandrel of the British Library's Universal Cylinder Player [Photograph], original source: British Library

Studer motor, that you mount the cylinder on with a REVOX parallel tracking tonearm that moves along the top of it with a stylus. The player also has a digital readout in terms of speed and can be adjusted between 50 to 320 RPM, which is critical when working with non-commercial cylinders where replay speed is non-standardised. For example, commercial cylinders were standardised at around 160 RPM, but the non-commercial types can vary quite wildly from less than 100 RPM to well over 200 RPM.

We therefore have to try to get the correct playback speed which often requires a lot of research and sometimes a subjective approach as there may be no reference point. For example, as the content is mostly music from other cultures around the world that people may not be familiar with, if I play it at one speed it could sound like a woman singing and another speed could sound like a man, so you need to do some research. Whatever decisions around playback speed, I always document the speed and the reasons why. Sometimes I choose a playback speed by ear and other times there may be clues (says it is a female vocalist, pitch pipe on the recording with key written down, matching speeds across a series of cylinders, knowledge of recording machine). The reason I was primarily brought on board to do this type of work is my solid ethnographic background and ability to undertake the kind of contextual research necessary to produce accurate representations, as opposed to purely technical experience.

Our whole process involves careful listening, researching, trying to get contextual information to verify what little description we have of the cylinder matches what I'm actually hearing. For example, if it says a solo female voice and I'm hearing a bunch of men, I know that I have the wrong cylinder and there's a bit of homework done. A lot of that is kind of fun and exciting, but also quite time consuming.

Digitisation

My primary responsibility is getting the content off the cylinder as objectively as possible. I see subjective processing like declicking, decrackling and noise reduction as a subsequent stage from what we do. However, we do try to understand what the original EQ curve would have been so that we can reverse this objectively during the process. From the player we send the signal into a PRISM ADA-8XR analogue to digital convertor and then into a PC running the software program WaveLab. We capture the digital audio transfer at a bit depth and sample rate of 24bit 96kHz, as we believe that by capturing the audio along with all of the noise, crackling and clicks at a relatively high resolution, we are future proofing the file to be subjectively processed in the future. The final file then ends up as a 24bit 96kHz WAV type. The WAV becomes the one channel master file, and in the longer term that will take the place of the physical item because over time the original artefact will become unplayable whereas the digital file should preserve for a lot longer.

Shellac discs

Shellac discs (Will Prentice)

At the sound archive, our shellac disc collection is generally made up of commercial classical and music hall light entertainment recordings – what passed for popular at the time up until the 1950s. We also house some non-commercial limited run recordings featuring linguistic surveys primarily produced for research and academia. For example, we have a set of shellac recordings of the Parable of the Prodigal Son spoken in every Indian dialect, but the vast majority of the collection is popular music from that time.

The shellac disc is more predictable than the wax cylinder in some ways, but the industry was not standardised until around 1955 so therefore speeds and the groove shape and profile were not standardised either. Up until the early 1920s, cutting or recording styli were made by hand and generally filed down from a size 5 sewing needle, so the groove can vary tremendously, and you have to find a stylus to fit. Playback at the time was generally done with steel or fibre stylus (as opposed to gemstones today) and they needed to be changed regularly. This was due to shellac records being made of two-thirds rock dust (generally slate in the UK), hence being a hard surface that would wear down your stylus within the first two revolutions to fit the groove so as to deflect the stylus in that left right path as accurately as possible. It is this robustness of the shellac record, and the vinyl resurgence resulting in playback equipment being readily available, that does not make it a priority within the British Library (unless rare) for digitisation as we have far more vulnerable formats that degrade more quickly, wax cylinders for example, that take precedent.

Depending on the shape and condition of the groove, we use either a conical or elliptical shaped stylus for playback to determine how deep down within the groove we want to place the stylus. If the groove has been worn down by a steel stylus, it has more than likely damaged the bottom of it. In this instance it would be better to use a larger truncated elliptical stylus to play from the middle and upper sections of the groove, avoiding the damage below. Typically, we use a truncated elliptical stylus between 2.5 and 2.8 thousandths of an inch for shellac discs.

Another challenge for us working with shellac discs is determining the correct speed for playback. The Gramophone Company in the UK stated that 78 RPM was the standard they adhered to, but this was not always the case. Similarly in the USA, Victor was recording at 76 RPM to deliberately introduce temporal distortion in the replay chain when played back at 78 RPM and Columbia sometimes wrote 80 RPM on their labels. In France, Pathé produced 18-inch discs that played back at 120 RPM, so as you can see speed varies greatly. At the sound archive we always write down the speed we use for transfer and the reasons why. For example, if we are aware that a recording of a violin concerto is in the key of F major, and when played back at 78 RPM it is clearly in the key of F# major, we would slow down the playback to achieve a digital transfer in the key of F major.

A further consideration with shellac is the associated EQ curve. Equalisation was standardised in 1955 for vinyl (and also shellac although it had been virtually replaced by vinyl by this stage) by the Recording Industry Association of America (RIAA). Prior to 1925, shellac records were recorded acoustically (via a horn and diaphragm) so there is no standard EQ curve identifiable. After 1925 when the Western Electric company made the first licensed electrical disc cutting chain consisting of an electric microphone, amplifier and disc cutter working together, their Westrex EQ curve was documented and well known. Shellac discs that portray a small triangle in the runout groove denote a Westrex disc, and other symbols represent other EQ curves from different manufacturers. We generally create a digitised flat master as well as one with the

EQ curve applied (only to discs recorded electrically) to offer a representation of what the original engineer would have heard. This creates a choice of master file for someone who intends to apply subjective processing and/or remaster from. Although we currently apply EQ to match the original curve in the analogue domain, we are moving towards adding this digitally as the software is so accurate as well as non-destructive.

For our digitisation chain, we use a modified Technics SL-1200 turntable modified to play 78 RPM (along with the standard 33$^{1}/_{3}$ and 45 speeds), remove the turntable motor and therefore the likelihood of interference, and a fluid dampener added to minimise resonance in the tonearm. From there the signal is sent to either a Ted Kendall Front End or a VADLYD Elberg pre-amplifier (we have other nice ones as well) which allows a wide range of EQ curves to be dialled up and applied, and then the processed signal is digitised using a PRISM ADA-8XR analogue to digital convertor and then into a PC running the software programme WaveLab.

Shellac discs (Mike Dutton)

Remastering from shellac discs varies depending upon whether they were recorded acoustically or electrically, and these need to be analysed separately.

Acoustic recorded shellac discs

Acoustic recorded shellac discs vary greatly in terms of speeds and groove sizes and therefore pitch is affected significantly, although acoustic recording systems did improve before they were succeeded by electric ones. For playback, I use a range of styli from 0.4 to 2.1 thousandths of an inch (and everything in between) due to this variance. My technique is to try different stylus sizes, depending on the wear of the disc. If it is worn, I can go lower into the groove to dig out the sound. If it is not too worn, I can go higher and get a better signal to noise ratio within that spectrum. I tend to use higher stylus sizes because I generally achieve greater gain from the recordings, and they playback with less noise.

Regarding speed and pitch, there is a fantastic book from that era called *A Dictionary of Musical Themes* by Harold Barlow and Sam Morgenstern. In this book, it has the first line of each work notated so you can see what key it was originally in. Therefore, if you have a disc recording, say of a Verdi composition, you can use the book to see what key it is in. You can then play a note on a harmonica or other instrument and adjust the speed of the disc player to tune it to the original key. There are also other books for vocal and orchestral themes as well as one devoted to Enrico Caruso recordings called *Caruso on Records* by Aida Flavia Artsay. These resources are all extremely useful for remastering, as is the original score of the music.

The frequency response of acoustic recordings generally tapers off at around 8kHz, but sometimes there can still be some high frequencies in existence. Additionally, if the disc wax gets cold towards the centre (and this happens with electrical recordings as well), cold wax chatter may be present. This results from the cutter head not being able to move the wax and it emits a whistling sound between 8 and 10kHz, which needs to be filtered out.

Another issue to be aware of regarding playback is how these acoustically recorded shellacs were pressed. If the discs are offset or they were pressed with an offset stamper, this may make it difficult to ensure the disc spins centrally, resulting in significant wow and flutter. To try and combat this problem I use a file to make the hole in the centre of the record slightly larger and move the disc around within the central point of the turntable to centre it. Of course, this works well remastering from one-sided discs, but with a two-sided disc, I always playback the best side first for centring and then use the file to correct the other side, so I don't damage the good side.

Acoustically recorded discs also tend to have been played with steel styli which wears down the disc and you can end up with a lot of steel (or thorn) in the grooves. To remove the steel and any other dirt in the grooves, I play it first and then use a Keith Monks or Loricraft record cleaner which both have suction through a fine nylon thread to remove all the dirt. But sometimes I clean a disc using the vacuuming function on a VPI record cleaner.

However, through cleaning you can dig out excess rubbish which has become embedded in the groove through age, resulting in more noise, so cleaning a shellac for remastering is not a straightforward exercise. It may be better to play a shellac disc first, then clean it, and then play it back. This process is repeated, generally no more than three times, to produce the best results.

Electrically recorded shellac discs

Electrically recorded 78 RPM shellacs were first manufactured and distributed through HMV and Columbia companies in the UK around 1924/1925. HMV used a moving iron cutter which spun the turntable via a pulley system, as used on acoustic recordings, but their speed was more consistent than acoustic recordings. Columbia used a Westrex electrical cutter at first, which spun the turntable via a clockwork motor, with the speed between 78 or 80 RPM. An issue for the early Columbia discs was that the speed would increase towards the end of the record, as the clockwork motor slowed down. This resulted in speed and pitch differentials with these early electrical recordings. However, the frequency response Columbia achieved on these early recordings were a big improvement, 8 to 10kHz, which was greater than what could be played back on home disc players at the time. For example, there is a recording of Gounod's opera *Faust* from this period, along with some dance band records, which sound amazing when played back on systems that can capture those frequencies. To solve Columbia's clockwork motor speed problem, and the Western Electric patented equipment, they later amalgamated with HMV to form EMI and enlisted Alan Blumlein (notable English electronics engineer and inventor) to design a new moving coil cutting head that used an upgraded pulley system. When I remastered recordings at EMI from discs that were created using the Blumlein microphone and cutter, the frequency responses were superb for their time.

In the 1930s there was a strong demand for dance records which led to more labels opening up in the UK, Decca (established 1929) being one of them. They used an early cutting lathe, which they felt had a fairly poor cutting head, at the time. Another smaller British label, Crystalate, were enjoying tremendous success through the Woolworth's market for dance music shellac records and used a disc cutter designed and operated by employee Arthur Haddy. Decca amalgamated with Crystalate to grab a share of this lucrative dance music market and as a result, discs were cut using a

Neumann lathe and Jenkins and Dare equipment, which was a copy of USA Western Electrics. With the updated Arthur Haddy-designed cutting head, the frequency response started to increase further.

When remastering from acoustic or electrical shellac discs, the variance of orchestral pitch between countries also needs to be considered. For example, if I remaster a performance of Richard Strauss' *Don Juan*, recorded at HMV, this will have A pitched at 440Hz (A440). If I remaster a performance of the same work recorded for Polydor in Germany, it is likely this recording would follow European pitch and be around A441–442. Some people I have worked with have suggested that, in this instance, the European-pitched A441–442 recording should be 'corrected' to A440. However, A440 wasn't established in the UK until later. Furthermore, recordings exist of performances conducted by Sir Henry Wood at A439 which is slightly under today's pitch, so variance is an issue. This can present a problem for the listening experience, especially for those like me who possess good relative pitch. For example, I recorded one of Michael Nyman's piano film scores in Munich at A442. This was sharp to my ear, and it took me a while to become accustomed to it. Everything was played well; it is just at a higher pitch.

Another consideration is the American versus European 'sound' on the recordings on shellac 78s. As America was so heavily based on radio recordings from this era, a lot of classical recordings were simply made by the station. For example, Toscanini's radio broadcasts were often transferred directly to records while broadcast and therefore included compression and broadcast limiters that were used at the time to ensure the radio transmitter would not overload but still pick up the sounds from the studio floor. Furthermore, the microphones they mainly used at this time were large Ribbon RCA-type microphones and the sound is therefore always in front (so the audience didn't miss anything) and not necessarily the best quality. If you compare a Toscanini American radio broadcast recording with a UK one recorded at the Queens Hall on HMV, the difference is quite noticeable as on the UK recording, you get a beautiful bloom and lovely sound from the BBC Symphony Orchestra compared with the dry, hard Studio 8H NBC/RCA sound.

Of course, there were some wonderful and inventive studio recordings from America at the time. For example, conductor Leopold Stokowski and the Philadelphia Orchestra got a great sound in the 1930s through pioneering microphone techniques including using multiple microphones and close miking. This was mainly brought about through budgetary constraints and the need to use a smaller orchestra, but the sound was fantastic. Furthermore, some of those Decca recordings from Los Angeles of Bing Crosby were beautifully engineered and recorded. Additionally, because of the quality of the shellac used (less filler added) to make the records, the American discs were quieter in terms of playback noise. However, although there were some improvements made to the American sound, overall it was still inferior to what was being produced in Germany and the UK, in terms of presence and clarity. One contributory factor to the sound was the European adoption of early Neumann CMV bottle microphones. Therefore, when I am remastering American 78 shellac recordings from this period, I adjust them through EQ to compensate for their sound. However, American recordings did improve during the 1940s through CBS introducing 16-inch 33 RPM microgroove lacquers.

Another interesting development from this period was the increase in frequency range, up to 18 kHz and higher, which occurred during the Second World War. This was due to the need to track German submarines and hear their high-pitched motors. An early example of recordings capturing these high frequencies during the war was the National Philharmonic Orchestra, conducted by Sidney Beer, performing Tchaikovsky's *Symphony No. 5 in E minor*. They were able to capture these wonderful high frequencies through the Neumann type M series capsules on microphones they designed, the type that was later housed in the famous U47 microphone, and a new FFRR cutter made by Arthur Haddy which is now housed at the Science Museum, London. In fact, they were able to go well above 20kHz and I have had to cut some high frequencies from these Decca K Series 78s from 1944 to 1949 when remastering, which is quite incredible when you think how long ago they were recorded, and the frequency spectrum captured.

Digitisation

It is therefore not a straightforward exercise to digitise and remaster shellac 78 recordings. I use an EMT 950 professional turntable (ex-BBC type) which is solid and has been modified so I can turn the speeds up and down. In fact, the EMT has less rumble noise than a Neumann cutter, and you can pick up the 35–40Hz rumble of a Neumann. The EMT is fantastic – I choose the correct stylus, it picks up everything

FIGURE 1.2 EMT 950 turntable with a selection of Ortofon styli [Photograph], photo courtesy of Mike Dutton

and I can go at any speed I need to. I generally listen through a couple of times, mainly to centre-check the pitch, then I transfer it to a digital audio workstation (DAW). I use CEDAR to make any adjustments regarding declicking, decrackling and noise removal and I then edit it all together.

This process is a great improvement from when I started remastering classical 78 shellac recordings at EMI Studios. We first used the Thorn/EMI noise removal system, which was a collaboration with the British Library sound archive and EMI Laboratories, Hayes. The problem was that it not only removed the noise, but it pretty much removed everything else. It was a laborious task as it was not in real time. We would copy the files onto the hard drive for processing at 5.00pm and they would be ready the next morning. Therefore, there was limited flexibility, and you could not make any adjustments in real time. The Thorn/EMI system was replaced with a CEDAR Cambridge system and things vastly improved from there.

I also could not work out why Stanton styli and pickups were being used at EMI. They said you needed a stereo stylus to listen to the lateral and vertical sides of the groove in order to pick the better side, which is true, and there was a Packburn noise removal system that would reduce the peaks of the crackle as well, but I didn't like the transistor-type sound the Stanton system produced. As a teenager playing old records on earlier equipment, I knew it could sound much better than I was hearing so I tried a different approach. I noticed an extra arm on the turntable and was told there were SPU-series Ortofon pickups with moving coils available in the store. I connected these up and found there was an equaliser on the switch that was never used. They were following a process of using quad preamplifiers with 25 microseconds of roll off at 78 RPM, which was the standard EQ curve for HMV 78s. I used the Ortofon pickup system with the other curve crossover box, and the improvement in sound was amazing as we were getting the correct movement within the groove and such a rich sound from the moving coils. In fact, the Elgar Edition we did using the Ortofons received a *Gramophone* magazine award in October 1992. I still use SPU-series Ortofon pickups today.

Acetates/lacquers (Will Prentice)

Some people refer to this format as acetate, but at the sound archive we call them lacquers as there is no acetate present and they date from the 1930s. Early shellac recordings were cut onto a wax blank disc about 1½ inches thick and then from there would be electroplated and copies made. The lacquer disc replaced the thick wax disc as it was easier and produced a better result.

They generally consist of a metal substrate (it could be glass) and cellulose nitrate which has plasticisers in it so that it is both soft enough to cut but also hard enough to play. These plasticisers are made from either palmitic acids, camphor or castor oil and they exude a mouldy grey look on the surface of the disc, causing the top of the disc to shrink. This can lead the disc to crack and/or delaminate as the top surface area of the disc is shrinking and the bottom section is not. They can also be made from gelatine, particularly if they date from the Second World War, which is important to identify because if the gelatine lacquers get wet through cleaning, we destroy the disc immediately. Therefore, one of the reasons that lacquers are our highest priority in the sound archive is that they degrade quickly and unpredictably.

Lacquers have the advantage that if you only wanted to produce one disc, you could just cut it, and play it back on any disc player or gramophone of the time. For example, The Beatles' first demo recording was cut to acetate/lacquer so that they could take it around with them to play to people. Elvis Presley's first recording was also cut on acetate/lacquer for him to give his mother as a present. Our collection of lacquers tends to be from the 1930s to the 1960s and in fact, in the sound archive we have a couple of acetates of the band that predated The Animals. Tape replaced lacquers as a format in the 1960s although some current DJs still use them as dubplates. The process for digitisation of lacquers in the sound archive is the same as for shellacs and vinyl.

Vinyl (Will Prentice)

Vinyl tends to be more straightforward than shellac and lacquers as they were standardised for speed ($33^1/_3$, 45 and 78), size (7 inch, 10 inch and 12 inch), groove and EQ curve by the RIAA in 1955. The thickness of the disc was also standardised by the RIAA and is interesting as it is not entirely flat. It has a certain thickness at the label, thins out as you move away from the label and then thickens again as you move towards the outside of the disc. As the groove is standardised, we use a 0.7 thousandth of an inch elliptical stylus, and for very early vinyl we sometimes go up to a 1.1 thousandth of an inch conical stylus. There are a small handful of various equalisation curves that were developed before the standardisation (Columbia had one and I think Decca had one) so we would investigate these as required, but for most vinyl it is standard.

We clean vinyl (and shellac) to remove obstacles out of the way of the stylus to achieve better playback and use a Keith Monks record cleaning machine which is really good at shaking dirt out of grooves. We also sometimes give vinyl an ultrasonic bath that we suspend the disc in, carefully avoiding getting the label wet. Due to their standardisation and equipment being readily available, vinyl digitisation is a low priority unless it is a super rare recording.

One challenge we do have with vinyl is that unlike books, we do not have much legal deposit in the UK so nobody is obliged to send us anything and we can't afford to buy everything. We rely heavily on donations of stock from record companies so maintaining a healthy relationship there is critical for the sound archive. For example, until the recent 2009 Beatles vinyl remasters, it was quite difficult to track down and listen to a mono version of a Beatles' album, but the sound archive held them, and they were available to the public to listen to. They sound quite different from the stereo versions and well worth a listen. For some researchers, it is the visual aspects of the vinyl recordings, say a code carved into a runout groove or a particular printed sleeve. Regarding digitisation, we follow the same process we use for shellacs for lacquers, but it is obviously a little easier due to the standardisation.

With digital releases, of course there is no physical product produced. We have developed a system for harvesting file-based content and metadata whereas with permission from certain record labels, they give as a monthly bulk download of their releases, and we can automatically source the titles and artwork as well as create a catalogue for us. It is quite impressive.

Vinyl (Mike Dutton)

I have mastered from vinyl when there is no master tape available, and the same principle applies as for shellac discs in terms of playback and cleaning. I am, however, not a fan of the current vinyl resurgence as I know that a lot of LP reissues have been cut not from analogue masters, but digital ones. I totally understand people, in particular younger people, gravitating towards vinyl as it is a physical product that they can own, they can watch it go round and share it, and it is warmer through its very nature. However, with vinyl playback we are back to the issues of wow, flutter and rumble, and then we have to look at whether these new vinyl issues are perfectly centred and the quality of the metal work behind the pressings. And more than likely, people are using spherical styli which are going to wear the groove down really quickly. They might be fine for tracking an old 45 that was cut loud, but I feel you should never use anything less than an elliptical stylus.

I own Neumann cutting EQs and limiters, the very old ones, and a cutting engineer friend of mine said the beauty about this old Neumann gear is that you put a digital signal through it, and it sounds like analogue. And the reason he and I believe this, is that the sound is going through passive EQs, a transformer and coils, so naturally through this physicality and mechanical process it warms up digital sound, and the frequency boosts and cuts move in ½ dB steps, and it can be made to sound lovely. Human ears are analogue. Of course, some vinyl reissues are cut from analogue masters, but not all of them.

Commercial tape (Mike Dutton)

The main concern with remastering from tape is that they are degrading, not all, but a lot. For example, we recently had a problem with a Charles Gerhardt (American conductor) recording of *Star Wars* in Kingsway Hall on BASF tape, where oil was oozing from the reels. You had to continually hand-wipe down the reels as we played them (seven reels in total) to take the oil off. It took hours but we managed to salvage these master tapes. Now there is a process where we bake tape, put them in an oven on a low temperature to help rebind the tape so you can play the tape through in order to transfer to digital (you have less than a week to do this once it has been baked), but this is not always the most appropriate step to take, and we could not do this for the Gerhardt tapes. In fact, baking a tape can do the opposite in terms of binding and would have worsened this situation.

I call BASF sandpaper tape as it is one of the best old tapes out there, solid and strong, but even they are starting to go off and become brittle. Another good solid tape is EMI. However, my favourite tape is Scotch tape, mainly due to the frequency response and dynamic range it was able to capture. I have used and remastered from Scotch 202, 206 and 250 tapes and I was at the launch of Scotch's 256 tape brand (although I didn't like the bass fuzz inherent on the 202 stock). Scotch tape was excellent, especially compared to Ampex tape, which always sounded flat to me. When you bake a tape, you lose a little bit of something, particularly with Ampex. Furthermore, if the tension on the machine wasn't correct, the Ampex tape would slide around: not so good when you had to find the next chorus to drop in for an overdub. It also felt a bit oily even then. Scotch tape was excellent and we used it at Morgan Studios and for The Cure albums I worked on.

I work a lot these days with stereo tape masters. It is important to maintain and look after your tape machines to ensure optical playback. I demagnetise the heads regularly, often removing the headblock first, as badly magnetised heads will add distortion to the tape. Obviously, I apply this process away from the tapes. I also often clean the pinch rollers and other machine components as old tapes that shed oxide can leave residue.

I tend not to do too much to the tape recordings; my approach is to intervene as little as possible as tape generally sounds good anyway, and I respect the original engineer's balance. You just need to line the machines up correctly through the test tones on the tape. If the tape master does not have these test tones on it, I use a test tape. When I use a test tape and line up all of the frequencies in sequence, I start with a 1kHz tone and then move onto 10kHz to adjust the azimuth (the alignment of the head), and then the bass. We check the frequencies all along the line and readjust as necessary. I also use a phase meter to see and help align the azimuth correctly, which is critical for remastering. If you don't follow this step, when you select mono on your monitoring equipment, the recording can be out of phase and disappear, and sounds can become sibilant and lose reverb. You need to be careful during this stage, however, as you can find a false azimuth, where it is the peak before the peak based on the angle of the head.

You also need be aware that some early tape recordings, particularly those made on the BTR (British Tape Recorder) manufactured by EMI, had more high frequencies than were able to be replayed by the BTR's heads. When I worked at Abbey Road in the 1990s, I used the inhouse test tape to align the Studer machines but sometimes an MRL Brand, and I found that the higher frequencies were raised on tapes recorded on a BTR machine compared to those recorded on a Studer. This explains why I was cutting the higher frequencies when remastering from BTR-recorded tapes. I also found that when I wound down the higher frequencies on tapes with alignment tones recorded on a BTR, I got better and fuller results. There are also various types of recording curve: CCIR, NAB, AES, AME etc. There is also the question of what the original recordings were monitored on (Tannoy JBL Urei and many others), plus in the faults of the room they were mixed in.

Generally, when remastering it is better to play back the master tapes from the type of machine they were originally recorded on if available. Furthermore, when you have a Dolby encoded tape you should also try to play it back through the Dolby system used at the time. A lot of these tapes have had Dolby applied so if you line the machines up correctly there won't be much noise evident, and the noise that is there is generally inherent in the room or in the ambience of the recording.

There is a bit of difference between the very early Dolby A301 and the later Dolby A Cat 22. For example, when recently working on a remaster I had to phone an ex technical engineer employee of Trident Studio, as they hadn't put any test tones on the 2-track master tape, and I couldn't get it to play correctly through the Dolby A system I have. I asked him why the tape was so bright, and he said that they used to pull out some of the cards of the four bands in the early Dolby system to get a 'harshness and presence' on the master recording, to create that 'sound'. Additionally, I had some feedback from someone on a Facebook group complaining they could not hear the mandolin on this recording, suggesting that I had somehow remixed it? You can't remix a 2-track master tape and I could hear the mandolin and others could as well. This also raises the issue of ageing, in particular with remastering pop and rock n roll with the treble and bass emphasised to excite people and compensate for older ears losing sensitivity in these areas. So, whose ears do you trust?

In my opinion, the best rule of thumb is, when you listen to it, if it sounds alright to you then don't fuck with it because it probably is alright. The original engineer knew what they were doing otherwise they would not have got the job in the first place. And if you are going to add something to it, you're most likely going to upset somebody, even more if it's a famous recording that everyone loves. A modern trend places far too much attention on the mastering or remastering engineer as opposed to the engineer who recorded and mixed it in the first place.

I own a Studer tape machine which plays back all speeds and sizes so I can pretty much playback any size or speed combination. With 24-track tape, I generally transfer it to digital 24bit 192kHz or DSD/DXD as soon as I can after baking it (to secure the bindings and edit points), due to degradation. For example, when recording The Cure and others back in the day, the original tape machine would have the tape shunting between forward and reverse continuously, which wasn't a problem at the time, but due to degradation, now you have to be a little careful handling these 40- to 60-plus-year-old materials.

Non-commercial tape (Will Prentice)

Our collection predominantly consists of recordings made by researchers and/or private people. It is a really interesting collection as it includes private recordings of BBC broadcasts which the BBC did not capture. We tend to use Studer A807 and A810 model tape machines for playback which was a strategic decision for the sound archive (although we do have old REVOX and TEAC ¼-track machines for slow tapes). By focusing on these two models (as opposed to owning a vast array of brands and type) we are able to build expertise within the team around maintaining them as well as source interchangeable parts (these are becoming scarcer). These machines can generally handle the speed and size combinations we use which is predominantly ½-inch full track mono, and, of course, stereo. We do not hold much multi-track material, mainly for the reason we would have to make subjective decisions about how it was played back. We would have to mix down the multi-track and make all kinds of ethical and artistic decisions that are not ours to make.

One concern for us when playing back a tape is adjusting the azimuth. Now unlike professionally recorded studio tape where you would expect the azimuth to be well aligned, with tape recorded by amateurs, this most likely was not a priority (or potentially a known requirement). Through aligning the azimuth correctly, we are getting the best playback we can in terms of sound level and frequency range. We also have Dolby A and Dolby SR units for any Dolby noise reduction encoded tape, but this occurs rarely.

Of course, being recorded by amateurs, we have some interesting tape configurations. For example, we had trouble identifying the recording characteristics of a tape consisting of spoken announcements followed by musical segments. It turned out that the tape was a compilation of spoken monophonic segments made using a ¼-track recorder, alternating with musical monophonic excerpts made using a 1/2-track recorder, all on one tape with no splices between segments. Bizarre! And we could only positively identify what had been done by using a magnetic viewer, a very handy tool which shows the tape track configuration.

It is well documented that polyester-based tape does not age well, and that baking can temporarily harden the binders on the tape for playback. However, when you hear friction on the replay heads through the tape squealing, this can be a symptom of the binders being sticky or another problem that baking may not fix. For example, some Austrian archivists contacted BASF and AGFA tape manufacturers and identified that tape squealing can emit from a lack of lubrication, not a binder issue, and from there we sometimes relubricate tape with isopropyl alcohol because it evaporates quickly and leaves behind no residue. It is a horrible messy job. For acetate-based tapes from the 1940s to 1960s, these tend to degrade much more quickly than the polyester-based tape and become brittle and curl. You can usually identify acetate-based tape as you can see through it when held up to the light, and if it is degrading it gives off a smell of acetic acid, or vinegar.

We have a small collection of 1940s tapes recorded in the UK shortly after World War II, on a German-made Tonschreiber, one of the world's first tape machines. We believe somebody from the British Army got hold of one of these machines and the original tapes, in Germany, and brought it home for home recordings and personal use. They would have been one of the first people in the UK to have a tape machine and they recorded radio programmes that predate BBC broadcasts, so it is very important content, although poor quality recordings.

The Tonschreiber machines were specified metrically, not imperially. So instead of a ¼-inch tape travelling at 30 inches per second, it was described as a 6-millimetre tape running at 77 centimetres per second. It also had nine different recording speeds (9–120 centimetres per second) and this was predominantly for recording morse code at slow speeds and playing them back at high speeds so they could not be decoded by the enemy in real time (without a Tonschreiber). Furthermore, the Germans invented tape bias during the war and that radically improved the quality of sound recording on tape. However, the speed is unstable as it varies according to voltage as it has a tachometer for the capstan. This research and context are all relevant to help us optimise the playback of the tapes, as is knowing what type of tape they are. There were three types of tape available for the Tonschreiber at the time: Type C – acetate plastic backing with a ferric oxide coating; Type L – PVC backing with a coke side coating; Type LG – homogenised PVC backing with ferric oxide in it. Being able to identify what the key component of the tape is, PVC or acetate, allows us to apply rehydrating techniques which will optimise playback on our Studer. There exists a train of thought that you need to source the original tape machines for optimal playback, but even if we could source a working Tonschreiber from the 1940s, we will get a much better digital transfer from a modern well-maintained high specification machine.

We also have in our collection some interesting tapes that are unavailable elsewhere. For example, we recently received a lovely Pete Townshend demo tape. It is a ¼-inch tape on a 3-inch reel of him playing 'Pinball Wizard', just Pete and his guitar. It came from a journalist who was friends with Pete, and his estate very generously gave us his journalist collection that included this. It was kept in a plastic box that has all these cigarette burns all over it.

Commercial cassette tape (Mike Dutton)

I have only remastered from cassette tape on a couple of occasions. I really like this format, there is something about them. I remember getting my first Philips tape cassette recorder when I was young, and it came with a Phonogram demo cassette. I just loved it. To me it always sounded better with the Dolby button not turned on (Dolby out) as this tended to expand the recording which I liked.

When remastering you have to denoise them. You also have to adjust the azimuth (usually with a small azimuth screwdriver) to get the best quality playback. However, you need to be careful doing this because of the pitch and speed, as although they are standardised to run at $1^{7}/_{8}$ inches per second, this does tend to vary and you are likely to need further adjustment.

Non-commercial cassette tape (Will Prentice)

I find they are fairly easy to work with as a format as they generally all run at $1^{7}/_{8}$ inches per second, the track configuration is almost always identical, and are simple to play. There are four main tape configurations which are machine readable due to indentations on the tape shell, unless made prior to the late 1970s. They do degrade over time, and we have recently started to experience tape stickiness. Sometimes, the tape won't pass through the cassette well, so we open up the shells and apply some lubrication. I try everything before I consider baking cassettes, because often the problem may be friction caused by other things (poor quality shell, tape imperfectly slick). The main consideration for cassette tape playback, however, is azimuth as aligning this correctly can make a huge difference in sound quality.

The sound archive stores a lot of cassette tapes in its collection with the majority emanating from oral historians using portable tape recorders, like the Marantz CP430. The use of noise reduction is common, in particular dbx and Dolby B, so that is something we need to be aware of. Furthermore, sometimes you are working with multi-generational copies, so it may have been encoded with dbx on the original recording and then not encoded on a copy made from the original, so it can be quite challenging to optimise the playback. You can therefore end up with a horrible decision to make of choosing the version that sounds the least awful.

We use a Tascam 122 Mark III cassette tape machine for playback as it is robust and produces great results. We can adjust the azimuth through removing the tray on the front and accessing the azimuth screw with a crosshead screwdriver. We have implemented an azimuth modification to our machines so that they last longer. This modification entails replacing the original screw around the azimuth (threads very fine and break easily) with a hex headed one as the threading is less fine and they last a lot longer. Although these machines are good, they have a sensitive mechanism that protects itself by shutting off if it's encountering too much resistance from the tape. Other machines, like those made by Nakamichi, have much higher torque and can playback tapes the Tascam cannot. Additionally, the Nakamichi Dragon has an automatic azimuth control which chooses the correct azimuth for the tape, which is amazing. It has a sensor ahead of the replay head that it automatically adjusts the azimuth for optimum tape playback which is perfect for a tape where the azimuth is variable.

Other formats

Dictabelt (Will Prentice)

The Dictabelt is an analogue audio recording format, released in 1947 by the American Dictaphone Company, whereby a stylus cuts a groove into a PVC belt. It was predominantly used for recording spoken words quickly, and that in turn could be provided to secretaries to type up the words onto paper documents. Although other analogue dictation formats and technologies arose during this time, the Dictabelt became famous as being the recorder that captured the audio emanating from a radio on a police motorcycle of the assassination of President Kennedy of the USA in 1963. Furthermore, Nelson Mandela's famous speech from the Rivonia Trial in 1963 was also captured for the court stenographer on a Dictabelt recorder. We possess a couple of Dictabelt machines and digitised the Rivonia Trial which was an amazing and important project of real significance.

Tefifon (Will Prentice)

The Tefifon is a German groove-based audio playback and recording system first developed in the 1940s which cut a groove into a looped infinite segment of 35mil film, complete with sprockets. This format allowed over an hour of recording, which was phenomenal compared to the 20 minutes you get from a tape at the time. The sound archive has a couple of Tefifon machines and in 1969 was able to transfer Tefifon recordings of Arnold Baker onto ¼-track tape. As a team we had a recent discussion regarding digitising these recordings. Should we try and get a Tefifon machine up and running so as to digitise from the original format, or use the ¼-track transfer completed in 1969? How much time can we invest in this and what will the potential benefit be? At this stage, it is likely we will use the transferred ¼-inch tapes.

Quad eight-track cartridges (Mike Dutton)

Occasionally, I have had to remaster from eight-track cartridges when the quad masters have not been able to be located. I actually own a Pro Dolby B system 330 so I can play it through and adjust to get the correct decoding levels. I have also decoded recordings from CD-4 Quadradisc and remastered from these for release.

Remastering from these formats

Will Prentice

We don't remaster at the British Library sound archive. We offer an audio preservation service. Our decision making is therefore objective and based around how best to use the equipment and techniques at our disposal to create the most accurate and best sounding digital copy of the original music artefact. We view remastering as a subsequent stage that is comprised of subjective decision making which occurs externally, and after we have completed our work.

For really challenging materials and we might want to make a kind of temporary subjective path just so we can get a little closer to what we think the content is, to ensure that we've identified the content and we're doing the right thing by it. I think the fact that we're literally saving the audio, sometimes we're the first people to hear it in over a century, it can feel a little bit like opening a wooden chest. Am I going to find treasure? Am I going to find junk? Whatever it is, it's going to be precious and important to somebody, especially if it's an ethnographic record from an under-represented or extinct culture. If we preserved it with due diligence, cared for it as well as we can, and taken all the steps do that, then we've done our job and I'm happy for it to be passed on to someone else, who knows and cares about the next stage of the process.

Mike Dutton

To be a remastering engineer, you have to be a recording engineer, as it is this experience and practical knowledge that provides an insight into mastering and remastering. For example, when denoising a recording it is easy to firstly conclude that the noise is coming from the format (disc, tape) where in fact it could be compressor noise and/or studio ambience. As a recording engineer you can identify the compressor noise from the vocal microphone and/or the ambience of the hall you are working in, and when remastering you need to ensure you are not removing something from the original recording. I do a lot of recording at Watford Town Hall, now called Watford Colosseum, which is one of the best acoustic spaces in the world (Maria Callas, classic Mercury LSO Recordings, Harry Potter film scores etc.), so I understand the ambience and reverb time, and this informs part of my remastering practice. Apart from Abbey Road, Air Studios and a few others in the USA, I am the only one of last few recording engineers to record music completely in the analogue domain up until the machine stage in DSD or DXD. In Watford, I have used over 60 channels comprising many valve microphones running long multicores through to an analogue Cadac mixer and the recordings I make sound great to me done this way. I try to keep the golden age of recording alive.

A way to remove noise is to take a noise sample before or at the end of the audio signal from the source, and then you can be confident using this sample in CEDAR that you are not removing sounds inherent with the recording. For example, if you have cold wax chatter at the end of a disc, you can take the noise sample at the end and apply this in CEDAR to filter out the whistling noise. Furthermore, you could declick, decrackle and denoise in real time using CEDAR NR4 and NR5 broadband spectrum denoising programs. It is amazing and works like a super noise gate. There are various other systems for cleaning up noise, but I still feel today that the CEDAR plug-ins and the ReTouch plug-in are still the best for restoration processes.

It's interesting remastering from shellac records and how audiences respond. Some people form what I call the 'scratch and crackle brigade' who believe digital remasters should contain all of the original noise associated with the shellac format, which of course worsens through poor maintenance and excessive playing over the years. I disagree with this view as the noise is only there because it was pressed on bad shellac and has no relationship to the recording itself. Of course, denoising will always remove a small part of the recording, be it room ambience or timbre, but there are ways of adding this back in and you have to be clever how you go about this. For example, adding a little reverb and getting the decay time to match the tempo of the recording can improve the room ambience. However, I have heard reverb added to the noise on some recordings and it is awful, like a horrible audio soup. So, I intervene when I need to, and I leave things alone when they sound great.

There is a belief that unprocessed remastered versions of shellac 78s sound 'brighter'. This is in part due to the filler makeup of the shellac disc which creates a natural treble brightness. If you compare HMV and RCA recordings from the same period, you'll find the RCA sound is slightly smoother and duller as it doesn't have the same amount of filler as it did in the UK. In India and Australia, there was even less filler in their records. Once these recordings have been denoised, these frequency differences are more exposed. Another challenge is when you have to join together recordings from different acoustic spaces. For example, Sir Thomas Beecham would record sections of a Symphony's Movement in one venue as well as another venue in the 1930s (Kingsway Hall, Abbey Road Studio 1 etc) or with later recordings in different countries (London, Paris etc.).

Sometimes the original master tapes or discs have been lost over time and you are required to create a remaster. For example, I was working on a project and the artist had lost their master tape. I manged to get hold of the artist's Japanese vinyl LP copy, which had never been played, and remastered from that. I played it on my EMT turntable with a Shibata stylus, transferred this into digital where I declicked it, it was released, and no one ever knew.

There was an article in the *Guardian* newspaper years ago that described what I do is to 'try to make the sound of yesterday, but for today' and to recreate listening to shellac 78s on a CD player at home. To do this, I enhance the room ambience where required but not necessarily adding compression as the recordings already have cutting limiters applied to the sound. There may be the occasional vocal peak from a dance band or a trumpet or an orchestral peak that needs attention, but you can just go in and edit that transient. I do believe you should try and make it sound better if there is nothing generationally or artistically lost from it. Besides, they can always go back to the original shellac recording. The other thing is the audience for 78s is and has probably declined. How many people are going to want to listen, say in 10–15 years' time, to Caruso, Gracie Fields or 'The Post Horn Galop'?

When I remaster rock music, and I come from a rock background, I respect what the original engineer has done. In the studio, I was hearing this great sound from say Brand X and the John Hiseman's Colosseum and bands like that. But on the LPs, it was half mono as the stereo was not 100% left to right because there is some crosstalk added at the lower end. To hold the bass, you have to mono it using an EE (an elliptical filter) to hold the styli in the groove, so the stylus didn't jump out of it. When I was mixing at the time, I made sure not to overdo the bass, and that artists were happy with what they heard as that was your job. Since CDs have been around, this problem has ceased to exist as you could go much 'louder' and put more energy into the lower frequencies, which is now synonymous with modern digital recordings since the late 1980s.

Some remastering engineers, working with analogue recordings predating CDs feel the need to make them louder and with more bass so that they sit more comfortably alongside modern recordings. To me, there is nothing worse than digital EQ, because you're bringing up louder nasty artefacts. If you're adding 5–6 dB at 15K and/or 4–5dB at 60Hz, the music becomes uncomfortable to listen to as you have added too many high or low frequencies and you therefore change the balance of the music. When they remaster bands now like The Beatles, Jethro Tull and others, that's done completely differently as they're taking a new look at it, to refresh, repackage and recopyright it mainly to preserve longevity and copyright, not necessarily to rebalance the original's recording (although this has been the work of Giles Martin and Sam Okell with the surviving Beatles' blessing in recent years – see Chapter 2). They generally have that available in the box set anyway. However, we have Dolby Atmos and 360 Reality Audio and other futuristic playback systems to come and still yet to be utilised so it will be interesting to see how that develops

When I was remastering The Guess Who, there was criticism that my remaster did not sound as bright as the remasters from America, and mine was flat and sounded boring. I simply did not add high frequencies. I decided to leave it as warm as the original record had sounded, and as the band and engineers originally intended, and ultimately you end up with a better product which is comfortable for the listener. Furthermore, adding too much brightness and high end can be tiring to listen to. There was a remaster done of 'Year of the Cat' by Al Stewart, originally recorded and produced by Alan Parsons, where there was a lot of complaints on internet forums regarding it being too bright and tiring to listen to. Remasters should be comfortable to listen to and draw you in.

I've survived in this industry for a long time and have plenty of customers so I must be doing something right. I think you need to listen to recordings as an engineer first and foremost, to understand the recordings and how and where they were made, and understand and have experience of studio noises that are inherent with the original recordings, and then remaster from there. You can learn from a book, but you can't beat hands-on engineering experience.

Cultural heritage (Will Prentice)

We try to use high performing modern machines (Studer as opposed to Tonschreiber) for our digitisation process, as we are trying to get the most accurate and best sounding transfer we can from our collection of artefacts though objective decision making. We do not remaster in a sense; we capture and preserve audio. But in the case of remastering say Elton John's *Goodbye Yellow Brick Road* album, I assume the engineers want to make it sound really good through applying many techniques and make subjective choices. However, I think it would make sense when working on a commercial remaster like Elton John to create two versions: a connoisseur version for listening to in quiet spaces with a large and lush dynamic range, and a 'meat and two veg' heavily compressed version to listen to on the go, earbuds on a bus, and to combat all the associated day-to-day noise. I think there is room for both. I am not a Puritan about equipment, but I am about sound quality.

With regards to future generations potentially replacing original musical artefacts with remastered versions, and viewing the remasters as the original version, this doesn't apply to me, but it does have an impact. I think it is a human characteristic to believe whatever we are doing now will last forever and be the definitive version. I imagine if I remastered or re-edited something today, I would feel it was the definitive version. But then you go forward ten years, and you realise you didn't because taste, the psychoacoustic evolution of what we expect, technology and what people want from music has all changed. Everybody thinks what they've done now is forever, and they're always wrong.

You could argue what is the definitive version of Miles Davis's *Kind of Blue* album. Is it the original mono pressing or the original stereo pressing? Is it the subsequent stereo pressing? Is it the remastered version? Is it the CD? Is it the Blu-ray? Is it the 5.1 version? You could argue maybe it's the job of an archive to have about half a dozen of these different versions.

I researched on Discogs online and at the time there were around 49 different versions of *Kind of Blue* issued in Europe that we conceivably might have in our collection. I looked through our collection and found at least 17 copies. If I had to digitise the analogue copies and turn them into files, where would I start and how many would I do? But you know that was just me playing around. I don't know if that is official British Library policy, but it was an interesting thought experiment.

When you hear a tape being played, you're never only listening to the tape, you're listening to the relationship between the tape and the tape machine. Even though you're listening to something pre-recorded, you're listening to a live event of that tape in that machine at that time. Now, if the machine is well maintained, you shouldn't hear any difference on the multiple times you play it, but there will be subtle differences so it's useful to keep that in mind. If you place a CD into an analysing device and read the error rates coming off it, play the CD again and the errors will be subtly different because it was a different experience the player went through. So, it is never about one element, it is about the relationship between artefact and player, artist and audience.

Cultural heritage (Mike Dutton)

The way I view digitally remastering from original formats is that you have to make them sound as best as possible because of the degradation of the sound from the artefact and the state of the artefact itself. Unless you try to enhance the sound now, add a little something to make it fuller, you may not get another chance to. You have to do what you think is right and remember that you cannot please all of the people all of the time. There is also the issue of what digital format to transfer and/or remaster to, as I have seen many digital formats come and go. What will be the correct digital format in years from now?

Regarding whether remasters become the new definitive version, I don't think there is one. In fact, over time I don't think there's going to be a definitive version of anything as everything is relative to time, situation and the state of the material. There is always going to be the original version, the first recording and release, and then subsequent versions from that point. I totally understand where Malcolm Davis, the ex-EMI/Apple/Pye/PRT cutting engineer, is coming from when he was remastering. He would always respect the original engineer's intentions. Malcolm used to say that you could always go back to the original source or a copy of the original source at the time (safety master tape) unless they were lost or unplayable. You need a good reference starting point.

EMI engineer Malcom Addey (Cliff Richard and The Shadows) was a brilliant engineer and said to me once 'when I'm listening to reproductions of these 45 records I used to make, I knew how I wanted them to sound and they sounded great that way, but when they remaster them now, they don't sound like that 45 any more'. I believe Malcolm's summation may have something to do with aural perception of the 45 RPM which refers to how you remember something sounding like, and how it actually did sound. For example, I remember my brother buying The Beatles' *White Album* and playing it on our old radio gramophone and I didn't think much of it. The replay equipment, in this case the radio gramophone, probably was not very good and replay equipment like that in general can never replicate the acoustics delivered to and listened in the studio. However, I also recall listening to The Beatles' *Abbey Road* on my fidelity system which had a crystal pickup and valve push pull amplifier. I thought it sounded great, especially the guitars at the top on 'Here Comes the Sun' and when the bass comes in – wow. But did it sound that great? From my aural memory, yes. So, I guess to get a realistic listening experience, you should play records on a record player from the same period because the sound is going to change as technology moves forward. Furthermore, as we age our ears and listening response changes, which is a whole other issue that goes back to aural perception.

Another issue is that for some people digital is too perfect, and as a result sounds a little flat. It does not have the imperfections and inconsistencies of analogue whereby components have different tolerance levels to the signal and hence create a colouration effect over the audio. People try to add colouration to sound through various digital plug-ins, but it cannot produce the same randomness of analogue warmth or of a mechanical process. Digital allows you to produce louder masters and remasters which can interfere with music's natural dynamic range. You should be able to play all types of music and it should come back at the correct dynamic range, which I do, as I know when playing the peak on an orchestral recording is going to be the same when playing as the peak on rock recordings. Not all just all loud!

I'm content to remaster most types of music, but I draw the line at thrash metal, hip-hop and the sample-based music of today. I have empathy for all the styles and genres I work on, which is very important.

When we talk about music's cultural heritage, I'm sure the British Library sound archive has guiding principles around audio preservation to a technical and/or industry standard. It would make sense. I don't adhere to any industry standard as I am me. I try to make the best reproduction I can from the original artefact, taking into consideration the original intent of the artists, arranger and the recording team, and the comfort of the listener.

2

REMASTERING THE BEATLES' *ABBEY ROAD*

Introduction

The album *Abbey Road* by The Beatles was originally produced and released on vinyl on 26 September 1969. Although it was the second last album released by the band (*Let It Be* was released in May 1970), it would signify the final recording project for John Lennon, Paul McCartney, George Harrison and Ringo Starr as a unit. It was hailed with critical acclaim with producer George Martin citing it as his favourite of all The Beatles' albums (Kehew & Ryan, 2006). Taking the title from the road outside EMI recording studio (as it was known at the time) where they had been entrenched since 1962, the iconic cover photo of the band walking across the pedestrian crossing away from the studio was a strong message, according to balance engineer Geoff Emerick, that this would be their last album.

> I didn't think it would be the last one. And nothing was said to indicate that, at least not to me. As far as I understood it, we'd be working on another record in the new studio I was building at Apple. The band was getting along better. The mood wasn't bubbly and fun all the time, but it was a hell of a lot better than during the previous year. The only hint they gave me or anybody was on the album cover, where they're walking across the street. For people who don't know the geography, they're actually walking away from the EMI Studios or Abbey Road, as everybody knows it now. This was intentional on their part – they didn't want to be seen as walking toward the studio. When I saw that photo, I did think to myself, 'They're sending a message'.
>
> *(Bosso, 2022)*

Towards the latter stages of recording the album, the band worked on a selection of potential album titles. One suggestion was to call the album 'Everest', based upon the favoured brand of cigarettes smoked by Emerick (Bosso, 2022). McCartney supported this claim, and in a 1989 interview he mentioned that Everest might be good as it was 'big and expansive', but the band didn't like it in the end as 'you can't name an album after a ciggie packet!' (Beatles Interview Database, n.d.). According to Emerick, it was Ringo Starr who

DOI: 10.4324/9781003177760-3

came up with the idea of shooting the cover outside as the band weren't keen to trek to the top of Mount Everest to shoot the photo (Bosso, 2022).

Originally released on vinyl, the album has throughout the years been released on various formats including cassette tape, 7-inch reel-to-reel tape, eight-track cartridge, Mobile Fidelity Sound Lab (MFSL) half-speed vinyl and cassette, CD, Blu-ray and mp3 (Discogs, 2022). This chapter will examine the production techniques used to create this original masterpiece in 1969 as well as the remastering processes applied to the 1987 and world-renowned 2009 remasters. Furthermore, perceptions of sonic differences between the 1969 vinyl, 1987 CD and 2009 CD releases will be explored through comparative listening and digital audio analysis.

EMI (Abbey Road) Studios

Although The Beatles would make some preliminary recordings for the album at London's Trident and Olympic Studios, most of the sessions for *Abbey Road* were undertaken at EMI Studios, as it was known prior to this album, which still resides at Number 3, Abbey Road, St John's Wood in north-west London. Celebrating its 90th anniversary in 2022, this studio is still one of the world's largest purpose-built recording studios (Bieger, 2012). Its rich musical heritage includes recordings by a diverse collection of world-renowned artists including The Beatles, Pink Floyd, U2, ABBA, Oasis, Ella Fitzgerald, Fats Waller, Yehudi Menuhin, Glen Miller as well as numerous orchestral recordings, and soundtracks for films such as *Star Wars*, *Lord of the Rings* and *Harry Potter*. Furthermore, the building received a Grade II heritage listing in 2010 protecting it from being demolished or altered (around the time EMI reportedly considered putting the studios up for sale), as well as the famous pedestrian crossing outside the front, which is rare as grade listing is usually reserved strictly for buildings only (Bieger, 2012).

The building consists of three main large acoustically treated recording spaces: Studio One (predominantly used for orchestral recordings), Studio Two (where most of The Beatles' back catalogue was recorded) and Studio Three. Upstairs there is also a collection of mastering suites and the Penthouse mixing and recording space, and on the bottom level is a canteen and bar area. Although the studio spaces are equipped with current gear to produce modern recordings, it also houses a vast collection of vintage equipment that is still in use and literally stored throughout the building in corridors, made famous through association with The Beatles and other recording artists and often built inhouse by the EMI technicians.

This equipment includes EMI-built REDD and TG consoles, BTR tape machines and RS124 compressors as well as Fairchild 660 compressor/limiters and a large vintage microphone collection (Bieger, 2012). According to Abbey Road's technical engineer Lester Smith, the microphone collection numbers over 800 and includes classic Neumann U47, U48, U67, M49, M50 and KM54 models as well as rare types built inhouse (Lester Smith, personal communication, 21 April 2022).

Recording *Abbey Road* in 1969

The 2006 book *Recording The Beatles: The Studio Equipment and Techniques Used to Create Their Classic Albums* by Brian Kehew and Kevin Ryan provides a comprehensive and detailed account of the studio personnel, equipment and techniques used to produce all of The Beatles' studio albums, including *Abbey Road*, and I strongly recommend this for further reading and details surrounding this phenomenon. According to Kehew and Ryan (2006), the production credits and equipment used to record *Abbey Road* was as follows:

Primary Balance Engineers
Glyn Jones, Jeff Jarratt, Phil McDonald, Geoff Emerick

Tape Ops
Richard Lush, Nick Webb, Alan Parsons, Neil Richmond, John Kurlander, Chris Blair, John Barrett, Roger Ferris

Recording Console
Studio One: REDD.37
Studio Two: TG12345
Studio Three: REDD5.1

Tracking/Tape Machines
3M eight-track
Studer J-37 four-track

Mixdown Machines
MONO: ¼″ EMI BTR2
STEREO: ¼″ EMI BTR3

Primary Outboard Gear
RS 124 compressor
Fairchild 660 compressor/limiter
RS 127 Presence Box

Effects
Echo chamber, Repeat Echo, double-tracking, half-speed recording. frequency control, ADT; flanging, Leslie speaker, tape loops

Microphones

Drums
Overhead: AKG Dl9c
Under Snare: Neumann KM56
Bass Drum: AKG D20 and Sony C38A
Toms: AKG Dl9c
Hi-Hat: AKG Dl9c

Bass
AKG C12 and DI box

Electric/Acoustic Guitar
Neumann U67, KMS4 and AKG D 19c

Vocals
Neumann U47, U48, KM54, KM56

Piano
Neumann U67, AKG D19c or C12

(Kehew & Ryan, 2006)

Initial reservations

There were initial reservations within the band as well as amongst EMI studio personnel when they assembled in February 1969 to begin working on the *Abbey Road* album. The band had only the previous month shelved their recordings for their Get Back project with no firm plan to complete or release the album (to be released in 1970 as *Let It Be*), and the accompanying film captured numerous arguments and walkouts amongst the band, depicting a fractured group with unity at an all-time low (Golsen, 2021). According to balance engineer Geoff Emerick, who had worked on previous Beatles' projects, tensions within the band had surfaced before during recording of *The Beatles* (known widely as the *White Album*) during which time he quit.

> 'The group was disintegrating before my eyes,' says Emerick. 'It was ugly, like watching a divorce between four people. After a while, I had to get out.' 'Oh, it was a nightmare. I was becoming physically sick just thinking of going to the studio each night. I used to love working with the band. By that point, I dreaded it. Getting out was the only thing I could do.'
>
> *(Bosso, 2022)*

However, Emerick would be swayed to join up with Martin again to record The Beatles' *Abbey Road* album. The band had decided to drop their 'no-overdub' ethos they followed for the Get Back project and were keen to return to familiar surroundings with familiar faces.

> 'After *Let It Be*, which I understand was not very pleasant for anybody, Paul was very keen to make a record the way the band used to. He wanted George Martin and I behind the console and everybody working together. He said things would be better than what they had been.' 'Yes, I did take him at his word. And John said the same thing to George Martin. In the back of my head I might have had some reservations, like, "Well, we'll see …" But I was surprised and pleased at how everybody got along.'
>
> *(Bosso, 2022)*

To herald the beginning of this significant project, EMI had recently built the TG12345 solid state mixing console for Studio Two in part to take full advantage of the 3M eight-track tape machines they had recently acquired. Apart from the shift from tube to solid state, the 24-input 8-output TG12345 boasted three times the amount of microphone inputs, had built in limiters/compressors on every channel and more sophisticated EQ controls compared to the pre-existing REDD.51 console it replaced and opened the door to more modern recording processes including recording in stereo (Womack, 2019).

However, Geoff Emerick was not initially thrilled with the new TG12345 console but acknowledged that it did make a difference to the studio workflow and overall sound of the album. As he states:

> 'Personally, I preferred the punchier sound we had gotten out of the old tube console and four-track recorder … It seemed like a step backward.' But in *Solid State*, he explains how that mellowness affected the recordings: 'With the luxury of eight tracks, each song was built up with layered overdubs, so the tonal quality of the backing track directly affected the sound we would craft for each overdub. Because the rhythm tracks were coming back off tape a little less forcefully, the overdubs—vocals, solos, and the like—were performed with less attitude. The end result was a kinder, gentler-sounding record—one that is sonically different from every other Beatles album.'
>
> *(Handley, 2019)*

Another technological addition to the project was the Modular Moog III synthesiser. Although the band had used the tape-based Mellotron sampler on the 1967 recording of 'Strawberry Fields Forever', this would be the first time the band had access to a true synthesiser. George Harrison had been impressed by the instrument during a recent visit to the USA and ordered one to be built for him in the UK. The modular set-up consisted of two keyboards, a ribbon controller, 901 series oscillators, 905 spring reverbs and a 984 matrix mixer and was used in the recordings of 'Maxwell's Silver Hammer', 'I Want You (She's So Heavy)', 'Here Comes the Sun' and 'Because' despite limited instruction and training available (Handley, 2019).

There were also some initial reservations over using Abbey Road Studios in general and the decision to record there was made partly through problems associated with their new studio at Apple, which required a completely new professional fit-out due to the well-documented problems associated with the original studios' build and design by Magic Alex. According to Geoff Emerick, who was employed by Apple to fix the studios, The Beatles band members felt 'incarcerated at EMI' and 'grew to truly hate the place', and they found it 'cold and quite uncomfortable and EMI slow to embrace new technologies' (Bosso, 2022).

Working at Abbey Road (Nick Webb)

I heard that Abbey Road was looking for people, this would have been around 1967, and I wrote them a letter, so I guess my timing was quite lucky. I'd played in a band since I was 12 and was interested in recording and had experience bouncing stuff from one tape machine to another and using microphones, which obviously helped me in the interview. During the interview, you were shown round the premises and told 'this is what you might be doing if you get the job' and everybody was wearing white coats and it looked serious. It was a good place to work as they were very inclusive; for example, our bookings were handled by a lady, so it wasn't all run by men, so it was quite good in that respect.

I began my career at Abbey Road in the tape library. This was an important introduction as I learnt about different types and sizes of tapes, how they were filed, how to find and retrieve them, and how to access the library after hours which all proved very helpful throughout my career. For example, I was working with The Beatles, and John and Paul suddenly decided they wanted to work on the track 'You Know My Name (Look up the Number)' which was being recorded around the Sergeant Pepper period. It was a four-track tape (we were working on eight-track by then) so I had to find it and quickly. I went to the tape library and discovered the tape was housed in a nearby squash court rented by Abbey Road for extra tape storage. I had to get the key and go down there, it was around midnight, and find this old tape. Anyway, I found it, we dug out a four-track player and plugged it in and set up the tape. John and Paul set up around one microphone and sang the vocals and we had to quickly drop them in and out on tracks as there was a lot of other stuff on the vocal track, careful not to erase anything. You were living on your nerves, making Chinagraph pencil mark-ups on the tape – drop it in then drop it out and they'd ask, 'Can we do it again?' and you could be doing this for hours.

From the tape library, I relocated to the technical department where I learnt how to solder and repair equipment. I was presented with a pile of headphones and repaired them. From there I moved to disc cutting, to the reference room where they created the reference lacquer cuts. It was quite involved as they had these old Scully lathes with no automation, so you had to do the whole process manually. Next, I went to work with David Bell, who was a classical editor, to learn how to edit tapes. David was a brilliant editor, and it was an honour to learn from him how to listen where to edit, mark up with a Chinagraph pencil and cut tape with brass scissors on a good angle to re-join the tape easily. To this day I prefer using scissors as opposed to an edit block as I could make a nice long angle which could make for a much smoother edit.

From editing I moved onto attending sessions as a tape op. My first session was working with Pink Floyd in Studio Two. I was shown how the clock worked for timing the length of songs and how the Studer J37 four-track tape machine worked. I learnt to listen to the engineer and the producer and to remain focused because if you lost concentration, you would not be much use to anyone. I remember working on a session for a Jake Thackray album and it was being recorded directly to stereo, so that was the master. The producer wanted to use a section from the first take and a section from the second take to complete the recording. Anyway, the engineer told me to do it as he was heading out for a cigarette. The producer told me what he wanted so I'm listening and marking up the tape. All the musicians were interested in what I was doing and crowded around me while I did this edit and my hands were shaking – remember I was still very green, and this was the master so if I stuffed it up, they would have to do the whole recording again. Talk about a baptism of fire – thankfully it worked out well and everyone was happy.

Abbey Road was also great because we had a solid team of technicians, including Francis Thompson who would thoroughly check all new equipment before it was allowed to leave his office. This gave you a great sense of confidence working with the equipment in the studio. EMI also built their own equipment and tapes at the factory in Hayes, including the BTR tape machines which were solid. And on the off-chance that a piece of equipment did fail during a session, the technicians would bring in a replacement and wheel away the faulty equipment for repair, so down time in the studio was kept to an absolute minimum.

Working with The Beatles (Nick Webb)

I worked on some sessions on the *White Album* and was asked if I wanted to work on *Abbey Road*. Vera Samwell used to manage the studio bookings for Abbey Road and said Nikki (she always called me Nikki) 'can you do some Beatles sessions on so and so?' And you know for us guys, you just sort of took your turn to be an assistant engineer with them. It was a real commitment, as The Beatles were the first band to sort of break with the traditional 2–3 hour booking slots of Abbey Road and record all hours during the night. You'd catch up with guys in the corridors that had recently been working with them and they looked tired. As for The Beatles, time seemed irrelevant and they could be long hours. I recall one session on *Abbey Road*, I was 19 and assistant engineering and Jeff Jarratt was about 21 and running the session, and I was thinking this was a bit odd – I expected one or more of the old hands to be there. But the good thing about The Beatles was they would be happy to work with a new guy if they seemed alright, and you must remember there were all these different faces coming in to work with The Beatles around this time like Phil McDonald, Ken Scott, Jeff Jarratt etc. My recollections of specific sessions are a little bit fuzzy given how long ago they took place. In general, I do recall a fair amount of tension around the band when they were all together which was not all that often. Working individually with the band was fine.

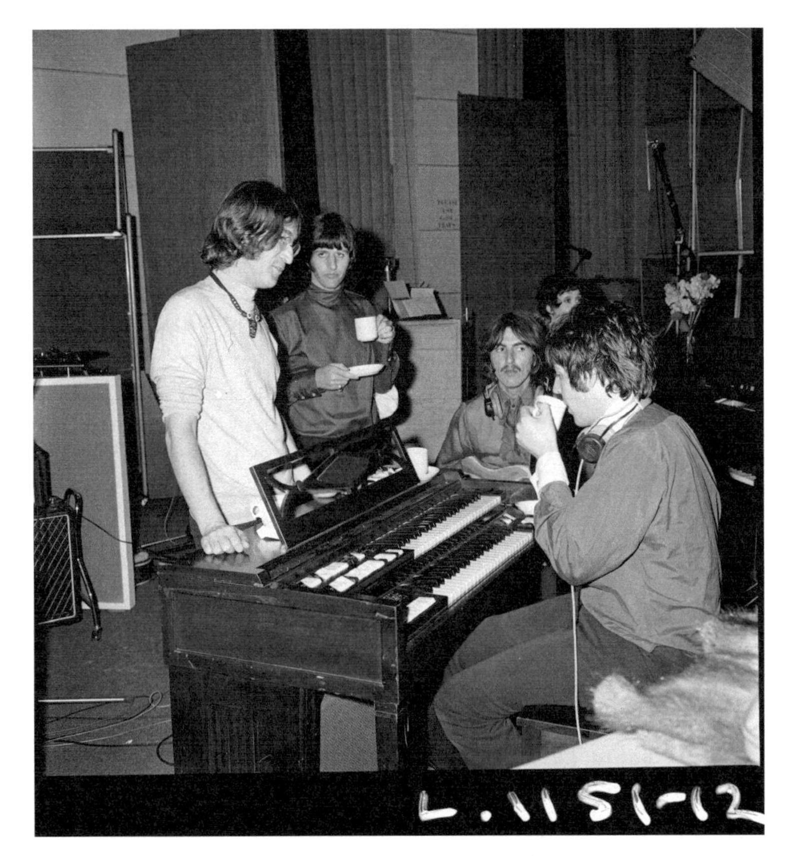

FIGURE 2.1 The Beatles in the studio [Photograph], received from *The Beatles Book Photo Library*

I think Paul was very much the driver of the band and very much a decision maker, as is obvious from what I've gleaned from the *Sgt. Pepper* album, you know in terms of leading the direction of the band. I remember seeing a lot of Paul because he would often come in early to sort things out. We spent some afternoons going through all the stuff that was recorded for the *Let It Be* album as nobody decided on what versions were good, bad and the best. So, Paul came in before everybody else to play through all the multi-tracks to try and sort them out into some kind of order. I also remember working as a tape op on a session for 'Come and Get It' which he wrote for Badfinger and he played all of the instruments on the demo, he was amazing. He then made scotch and cokes in the control room. 'Would you like a scotch and coke Nick? Yes please!' He is very inclusive, pleasant, and probably the easiest one of the band to get on with.

John was so unpredictable, but he could be a lot of fun. We were working on his *Live in Toronto* album (I think Mal Evans was the producer – he was a lovely guy) and John wanted some effect on the vocal and came and showed me where he wanted it placed as I was working in a little room off the studio looking into the control room. He said, 'I feel bloody daft.' He'd just ripped his tight leather trousers and said 'look at this' and his arse was hanging out. He saw it as this big joke! He could be cool and very funny, but on other occasions he could be quite cutting. George was very quiet as was Ringo, to me anyway, that's how I remember them.

Sometimes you just felt as though you're just there like one of the tape machines. You know what I mean, because that's what it was, you were working as an operative. The guy at the desk was just an operative as well. You had to be aware what was going on and on your best behaviour, but it's a long time ago.

Working with George Martin (Nick Webb)

He was a lovely super guy and I worked with him on several things with different artists. George was a great musician who read music and provided scores for musicians to play. I recall attending an Abbey Road event shortly before I left, and George was there with Chris Thomas and they both remembered who I was (I hadn't seen either of them for years). George was professional and treated everybody with respect; even working as a tape op for him, he didn't treat you as a lackey, he treated you as an equal.

Recording specific tracks

'Octopus's Garden'

In a 1981 interview, Starr mentioned that he wrote this song after a conversation he had with a ship's captain, who told him that octopuses move around the sea collecting shiny objects and building gardens, which Starr thought was marvellous as he wanted to hide under the sea around that time (Beatles Interview Database, n.d.). Harrison said in 1969 that he really liked the song, and that Ringo most likely wrote it while playing piano at home as he got bored just playing drums (Beatles Interview Database, n.d.). According to Geoff Emerick, Starr was shy presenting his compositions to the rest of the band and the fact that Harrison was keen to work on it was helpful and that it was a fun song to work on. As he states:

> There's this fun bit where you hear bubbles, as if you're underwater. Ringo tried blowing bubbles into a glass of water which we mic'd very close. In the end, I recorded his vocals, fed them into in a compressor and triggered them with this pulse-like tone that created a wobbly, 'bubbly' sort of sound.
>
> *(Bosso, 2022)*

Nick Webb

I remember setting up for 'Octopus's Garden' and making sure everything was plugged in correctly and working, and Ringo walked in before everybody else had arrived, and bless him, he seemed a little low and troubled. And he just sat down and started talking to me and I didn't really know them that well as people, they were just the artist (EMI's biggest artist of course). He just opened up and said, 'I don't want to sing this song and they always want me to sing a song. I don't know about doing it on this one.' I was thinking what do I say? So, I ended up saying 'well you've always done one and you know people love it when you do and it's what's expected'. It was a very weird experience.

'I Want You (She's So Heavy)'

A simplistic John Lennon composition about his love for Yoko Ono which was criticised at the time. As Lennon states in a 1971 interview:

> Simplicity is evident in 'She's So Heavy'. In fact, a reviewer wrote 'He seems to have lost his talent for lyrics, it's so simple and boring.' When it gets down to it – when you're drowning, you don't say 'I would be incredibly pleased if someone would have the foresight to notice me drowning and come and help me, you just scream.'
>
> *(Beatles Interview Database, n.d.)*

According to Kehew and Ryan (2006), this song was the first Abbey Road track started and one of the last ones completed. The initial recording on 22 February took place at Trident Studios where 35 takes were recorded. Recording continued in April where John and George added multiple guitar tracks (deleting Billy Preston's organ part) and completed in August with Lennon adding white noise from the Moog as well as vocal overdubs.

Geoff Emerick found the song fascinating. As he says:

> It goes from hard rock to almost jazzy, bossa nova. Of course, there's the famous ride-out, the riff being repeated many times. George put some very intense Moog sounds down and Ringo played with a wind machine – the whole thing grew louder and louder till it got close to a breaking point. I thought the song was going

> to have a fade out, but suddenly John told me, 'Cut the tape.' I was apprehensive at first – we'd never done anything like that. 'Cut the tape?' But he was insistent, and he wound up being right. The track, and side one, ends in a very jarring way.
>
> *(Bosso, 2022)*

Nick Webb

I remember working on 'I Want You (She's So Heavy)' in Studio Two and George Harrison was putting his guitars parts on the bit towards the end of the song as it builds up. It appeared to me that George seemed at a loss on what to play and did not know where the song was heading. It was just this relentless build-up of guitars, and it was quite chaotic, bedlam really compared to other sessions where there seemed to be a clear approach and pathway, well that's what it sounded like from the multi-track which somewhat surprised me. But then you hear the finished track, and it is brilliant.

'Something'

A George Harrison composition, described by Lennon as 'the best track on the album' and by Paul as 'the best song he has written', was written during the production of the *White Album* (Beatles Interview Database, n.d.). As Harrison mentioned in a 1981 interview:

> *Something* was written on the piano while we were making the *White Album*. I had a break while Paul was doing some overdubbing, so I went into an empty studio and began to write. That's really all there is to it, except the middle took some time to sort out. It didn't go on the *White Album* because we'd already finished all the tracks.
>
> *(Beatles Interview Database, n.d.)*

Geoff Emerick described this song as a turning point for the band where John and Paul realised George had written something special and up to the standard of Lennon/McCartney compositions. As he states:

> George had a smugness on his face when he came in with this one, and rightly so – he knew it was absolutely brilliant. And for the first time, John and Paul knew that George had risen to their level. Paul started playing a bass line that was a little elaborate, and George told him, 'No, I want it simple.' Paul complied. There wasn't any disagreement about it, but I did think that such a thing would never have happened in years past. George telling Paul how to play the bass? Unthinkable! But this was George's baby, and everybody knew it was an instant classic.
>
> *(Bosso, 2022)*

As described by Kehew and Ryan (2006), the recording took place in May and that the song would eventually end up as Harrison's first A-side single for the band. This song would be recorded in several different locations with 36 takes of the backing track in Studio Three, overdub sessions for bass and lead guitar at Olympic Studios, and lead and double-tracked vocals recorded in Studio Two. Furthermore, the orchestral session took place in Studio One with mic lines patched through to the control room of Studio Two into the TG12345 console, with the engineers monitoring the orchestra on a black and white monitor via closed circuit TV. As they only had one track left for the orchestra overdubs, Harrison recorded the guitar solo live with the orchestra, a gamble according to Emerick, but one that paid off as Harrison did it in one take and 'it was beautiful' (Bosso, 2022).

Nick Webb

I remember the band working on the backing track for George's song 'Something'. John turned up quite late, dressed in a big black hat and black jacket, and started playing piano to it and we did take after take after take. I'm not sure that ended up being the final backing track, whether they didn't use the piano and/or still used the bass and drums. Often, they would have several goes at recording songs and it could go on take after take after take, and sometimes they came back to it later to decide – usually the songwriter so in this case it would have been George Harrison.

'The End'

A Paul McCartney tune which Lennon described as 'Paul's unfinished song' with the great philosophical line 'And in the end the love you take is equal to the love you make' (Beatles Interview Database, n.d.). In an interview in 1994, McCartney described the line as something that just came into his head and that he thought it would make a good ending for the album. As he states:

> I'm very proud to be in the band that did that song, and that thought those thoughts, and encouraged other people to think them to help them get through little problems here and there. So uhh … we done good!
>
> *(Beatles Interview Database, n.d.)*

'The End' was the last song recorded for *Abbey Road*, with recording beginning on 23 July, and is noteworthy for several reasons. It was the only Beatles recording to feature a drum solo and have drums fill the entire stereophonic space and it also featured guitar solos recorded simultaneously by John, Paul and George together in the studio at the same time (Kehew & Ryan, 2006). According to Paul, Ringo hated drum solos and the idea of doing one. As he stated in a 1988 interview:

> Ringo would never do drum solos. He hated drummers who did lengthy drum solos. We all did. And when he joined The Beatles we said, 'Ah, what about drum solos then?' and he said, 'I hate 'em!' We said, 'Great! We love you!' And so, he would never do them. But because of this medley I said, 'Well, a token solo?' and he really dug his heels in and didn't want to do it. But after a little bit of gentle persuasion I said, '… it wouldn't be Buddy Rich gone mad', because I think that's what he didn't want to do … anyway, we came to this compromise, it was a kind of a solo. I don't think he's done one since.
>
> *(Beatles Interview Database, n.d.)*

George Martin was happy to follow the marketplace and turn his attention to producing stereo recordings (all previous Beatles recordings were originally produced as mono releases) as he appreciated the evolving multi-track recording technology and how this could benefit stereo recordings (Womack, 2019). The extra inputs of the TG12345 console in combination with the 3M eight-track tape machine, allowed for the drums to be recorded in stereo. Emerick could now dedicate two tracks to the drums (as opposed to one when using the four-track) and his drum solo recording for 'The End' was the only time they were recorded in true stereo. As Emerick explains:

> This was the first time I was able to record his [Ringo's] kit in stereo because we were using eight-track instead of four-track. Because of this, I had more mic inputs, so I could mic from underneath the toms, place more mics around the kit – the sound of his drums were finally captured in full.
>
> *(Bosso, 2022)*

Nick Webb on recording *Abbey Road*

I'm surprised and delighted that The Beatles thing hasn't gone away. There's been this longevity in their music because they were a phenomenon. They really did change the face of music and left us with lots of really good songs. There is a lot to be grateful for and I am pleased to see them still getting credit for it.

Mixing *Abbey Road* in 1969

The arrival of the 3M eight-track tape machine heralded a significant shift in mixing workflow at Abbey Road. Before this technological shift, earlier Beatle records were restricted to four-track (and a type of twin-track recorder before late 1963) which resulted in the engineers and tape ops having to undertake reduction mixes, bouncing down a number of tracks to another tape machine that had been hooked up, to free up tracks for more recording (Kehew & Ryan, 2006).These reduction mixes would introduce noise and generational loss of fidelity on instruments each time they occurred, so could be problematic. Internal bouncing, whereby you bounce down tracks to one track on the same machine was a better option in terms of noise, but with only four tracks in total available significantly limited how many tracks you could use for recording. With eight tracks now available, there were greater options available to reduce noise through limiting recordings to eight tracks and/or using internal bouncing whereby two or more tracks could be bounced down to one track on the same tape machine. However, there were some eight-track to eight-track (8T-8T) reduction mixes on *Abbey Road*, in particular 'I Want You (She's So Heavy)', where John Kurlander described the tape hiss on the quieter parts of the song as extremely audible (Kehew & Ryan, 2006).

Mixing 'The End'

Kehew and Ryan describe in detail the mixing of 'The End' which shows the complexity surrounding the task with the equipment available, particularly when you compare this to using a modern digital audio workstation (DAW) system. Here is a following abstract from their book:

> The song was remixed for stereo on 19 August. However, doing so revealed some problems. It was realised that the timing of the orchestra was noticeably off toward the end of the overdub; the last note came in well behind the beat. To remedy this, on 21 August the last section of the song was bounced over to a four-track tape, the backing track being reduced to a stereo pair on two of the tracks, and the two orchestra tracks being bounced over to the other two tracks. After some slight editing by Phil McDonald, it was then Tape Op Alan Parson's job to fly the orchestra into a new mix of the song. The original orchestral overdub on the eight-track tape was muted, so only the time-shifted orchestra on the four-track would be heard. (The backing track that had been bounced to the four-track gave Parsons a reference for syncing the tapes.) 'That end section was flown-in,' he says. 'I remember that I started the four-track machine just before the string line that follows "... is equal to the love". It took several attempts to get the timing right.' Once the ending section had been properly synced and mixed, it was edited into the final mix.
>
> It appears that the mix at this point still included the four measures of lead guitar that followed Ringo's drum solo and preceded the main guitar solo. It was decided to remove those bars and replace them with an edit piece that omitted the lead guitar. At the same time, the decision was made to lengthen the space between the drum solo and guitar solo. To accomplish this, Phil McDonald and Geoff Emerick created a new mix of the middle section, omitting the guitar solos completely. Several measures were then edited into the stereo mix before the guitars, extending the song significantly. Four days later, Emerick, Martin, McDonald, and Parsons would return to the song again for more editing. This time they removed some of this material, having apparently decided that too much space had been added during the previous session. The middle section of the final song therefore, is a patchwork of edits and duplicated measures.
>
> *(Kehew and Ryan, 2006)*

This tremendous insight into the editing and mixing of 'The End' clearly shows the complexity of producing the end product with the equipment available, particularly with associated time issues. According to Mark Lewisohn's detailed 1988 account of The Beatles' recording sessions, stereo remixes for songs on *Abbey Road* were conducted predominantly throughout the month of August 1969, culminating in the 20 August session in Studio Two where the final master tape for the whole album was compiled. At this point there were two variations on what would become the final released version; Side A and Side B were reversed: the medley was originally on Side A and the album finished with the severe cut on 'I Want You (She's So Heavy)' and 'Octopus's Garden' and 'Oh! Darling' were in reverse order (Lewisohn, 1988).

Mastering *Abbey Road* 1969

Abbey Road was mastered (or cut as it was referred to back then) at Apple Studios by former EMI engineer and veteran disc cutter Malcom Davies, from a master tape delivered by Geoff Emerick (Womack, 2019). One interesting development that happened during the cutting stage was the final placement of McCartney's song 'Her Majesty' on the album. In an article for *Variety* written by Steve Marinucci, *Abbey*

Road tape op John Kurlander described the event of how 'Her Majesty', originally placed before 'Polythene Pam', ended up as the final song on the album.

> 'This was the day that I found out that this whole thing was to be a medley,' he recalls. 'We joined it all together. It was about two or three o'clock in the morning and after a very long day, we played the whole thing through for the first time. Paul said, "Look, I don't think 'Her Majesty' works. So just cut it out," and he left and went home. So, it was my job to tidy up the housekeeping. And there was a piece of tape which was only 20 seconds long lying on the floor. There is an [EMI] rule that says if you remove something from a master tape, it has to go at the end, after a long piece of red leader tape. Everyone else had gone home, so I decided to just tag it on at the end. Then [long-time Beatles assistant] Mal Evans took the tape, and the next morning they had a reference acetate cut from it by Malcolm Davis, Apple's cutting engineer. Davis left 'Her Majesty' in, cut the reference of the medley, and brought it in the next day at lunchtime and played it through. And then this thing ['Her Majesty'] crashes in, because it still had the crossfade on,' Kurlander says. 'Paul was probably the most surprised, because his last word on the subject was "Just get rid of it."'
>
> *(Marinucci, 2019)*

All previous albums by The Beatles had been cut by Harry Moss at Abbey Road so this marked a significant change (Lewisohn, 1988). Although we do not have the details regarding the cutting of *Abbey Road* by Malcom Davies at Apple Studios, Malcom Davies is sadly no longer with us, Nick Webb describes below the likely mastering workflow Harry Moss (sadly no longer with us as well) used on all previous Beatles albums cut at EMI.

Mastering The Beatles prior to the Abbey Road *album (Nick Webb)*

Harry Moss mastered all The Beatles' albums except *Abbey Road*. Harry had come from the factory in Hayes, and he was the pop album cutting engineer. Even when I became a cutting engineer at Abbey Road, you were assigned to work on either singles, albums, reference cuts, classical or pop. There were different people in different rooms, all from the same department, doing these specific things. My understanding is that he was very much involved in their albums, and I know he had a challenge with the run out groove on *Sgt. Pepper's Lonely Hearts Club Band*. Apparently, that took forever. I'm not sure if the band ever got involved, they may have left George Martin to get on with it or not, they may have just sent Harry the tapes, it's hard to know.

The most likely workflow is that Harry would receive the master tape from either the Tape Library or the Progress Office and it would have been analogue ¼-inch mono or stereo on EMI tape (manufactured at our factory in Hayes). The finished format would be either 12-inch or 7-inch reference lacquers for the band and George Martin to listen to, and these would be a proximity of how the final vinyl disc would sound. However, lacquer reference cuts sound less noisy and a little warmer than vinyl pressings which have a firmer lower end. Once the reference lacquers had been approved, Harry would have cut onto a 14-inch lacquer for an album or a 10-inch lacquer for a single as per the reference instructions. These masters would not have been played (a single play can cause damage to the soft lacquer groove) and placed carefully into a safe tin, the same

FIGURE 2.2 Nick Webb

as the new lacquers were delivered in, and would be delivered by an EMI van driver to the factory for processing.

The main goal Harry would have been striving to achieve would be to get as much level on the disc as possible, a clean sound, without incurring problems from distortion and/or pickup stylus jumping on playback. There are many variables including the length of the side, whether it was an album or single, and as I understand, Paul was keen to have a loud bass sound, so all these issues would have confronted Harry. Additionally, there were set parameters as to how far a cut could go at the end of the disc to ensure auto-lift record players would play the record to the end, as well as the physicality of the disc and that it was prone to distortion closer to the centre due to the short distance the groove had to travel. All these technical issues had to be addressed or the master lacquers would be rejected at the processing plant.

Remastering *Abbey Road* 1987

The Beatles' official studio back catalogue, including *Abbey Road*, was released in batches throughout 1987; however, *Abbey Road* was previously released in 1983 in Japan via Toshiba-EMI Ltd but for a very short time before being removed from the market (Zaleski, 2017). There were no bonus tracks, alternate mixes or liner notes (unless liner notes appeared on the original album sleeve) included and for most of the catalogue, including *Abbey Road*, it was simple transfer from the original master tapes onto the new digital format (Belam, 2007). There were also no credits for who undertook the digital remastering process.

In an interview with *Record Collector* magazine, remastering engineer Mike Jarret described the process of remastering the collection. Extracts from the interview follow:

> There was an issue on whether to issue mono or stereo, or whether to remix. If you could hear the mono and stereo master tapes, you'd be in no doubt that the mono tapes sound better. I did check all the paperwork, and it appears that much more time was spent on the mono mixes, because back at that time very few people had stereo. In fact, I did prepare both mono and stereo digital master tapes for the first four albums. George Martin came in and give the OK for the mono tapes. The third and fourth albums were recorded in 4-track, and proper stereo mixes are obviously available; but even with these albums they were produced for a predominantly mono market. That's why they were issued on CD in mono.
>
> I prepared digital stereo master tapes for CD production from the original stereo tapes for *Help, Rubber Soul* and *Revolver.* George Martin came in to listen to the results, and he though the quality of the CDs would be improved if we went back to the 4-track tapes and remixed *Help* and *Rubber Soul.* I transferred the original 4-track tape onto a digital multi-track machine along with the original stereo mix, so George had a reference of what he had done originally. The remixes were then done from these digital tapes. *Revolver* was left alone as the stereo master sounded so good.
>
> With regards to digitally manipulating the masters, I referred to the original documentation to see what was done in the Sixties. I've had a couple of years' experience as a cutting engineer, so I knew what changed had to be made for purely technical reasons and what had been done for artistic reasons, so I hope I managed to get the best from the tapes. I even did experiments with running the master tapes through the machines that were in use at the time; but I found that the best sound came from using modern machines with 2-track heads. I guess if anyone purposely 'tampered' with the mixes, most Beatles fans would get annoyed. I have had to use noise gates and filters and so on some things, but we have kept their use to a minimum. The problem is that the more processing you use, the more you interfere with the original recording. People must accept that we're talking about a 19-year-old recording, and so it will clearly have hiss etc., that mightn't have been present on a 1987 recording.
>
> *(Doggett, 1987)*

The approach from George Martin and EMI to release the first four albums in mono as opposed to stereo was an interesting decision, especially when considering CD releases at the time and expectations of the audience were predominantly stereo. George Martin stated the following in an interview with Billboard in 1987:

> I was asked by EMI to give my opinion, and my advice was not to put them out the way they were doing. They were going to issue fake stereos – the bane

> of my life for the past two decades – and I said these were recorded in mono, so why do this?
>
> *(Zaleski, 2017)*

There has been criticism for the 1987 releases, particularly the remixes of *Help* and *Rubber Soul*, where a level of echo and reverb had been introduced (which were missing from the 1960s releases) and this was added through George Martin's approach to make them more modern by adding a 1980s stereo soundscape (The 1987 CD mixes, 2009). Martin described it as 'cleaning up the sound a bit' in an effort for the audience to hear it in a way that was not possible in the 1960s (Pacuta, 2020).

Remastering The Beatles (including *Abbey Road*) 2009

The Beatles' official studio album back catalogue was remastered and released in 2009 as either a stereo or mono boxed set, and available on CD, vinyl and a USB (which had the highest digital resolution at 24 bit 44.1kHz as opposed to the CDs which were 16 bit 44.1kHz). The engineering team at Abbey Road Studios, including Sam Okell, spent four years to create the 'definitive' digital version of arguably the most important contemporary music back catalogue in the world (Inglis, 2009).

> Faced with, on one hand, the demands of purists, and on the other, the expectations of modern listeners, the team chose to take two directions at once. For collectors and audiophiles, they created a box set comprising all the original mono versions of The Beatles' albums (less *Abbey Road*, which was not issued in mono, and *Yellow Submarine*, where the original mono was a straight fold-down from the stereo), which for the most part was as faithful as possible to the source. Simultaneously, they reworked the stereo catalogue for release in a second box set, and also as individual albums – again treating the material with respect, but not shying away from the application of modern technology, if it was felt that fidelity could be improved.
>
> *(Inglis, 2009)*

The decision to undertake the mammoth project was due to several reasons. Firstly, the existing CD remasters from 1987 were incomplete (some were mono and some were stereo but not all mixes were accounted for), they were inaccurate representations of the 1960s masters (*Help* and *Revolver* had been remixed) and audio technology had improved vastly between 1987 and 2009 which would ensure better audio quality and digital transfers (Inglis, 2009).

Working at Abbey Road Studios (Sam Okell)

I'd probably describe myself as a recording engineer and a mixer as opposed to a mastering engineer. I studied the Tonmeister course at the University of Surrey in Guildford and spent my placement year at Abbey Road editing classical recordings. After I completed my degree, I got a position with Abbey Road as a runner where I made the tea and cabled coils. From there I became a recording engineer and was just assisting some of the guys who were working in that world of remastering and remixing stuff. I was working with a guy called Paul Hicks and he was remixing a load of new 5.1 mixes for a Paul McCartney greatest hits record. I assisted and watched how people operated in the studios, that's how I learnt. For me, remastering is reappraising the finished mixes or the finished album in the current climate. Looking at what we could do to this with the tools available now to improvc, although this is a tricky word, to make it more relevant to the here and now.

Remastering The Beatles (Sam Okell)

I wasn't involved at the start of The Beatles' remastering project. I joined around midway through the project courtesy of my work on Paul McCartney's stuff which was happening at the same time. They already had a system in place, and it was working well. They had set up the transfer of tapes, restoration stage, and then the EQ stage so I just slotted into that really.

The Beatles' remastering project was a little bit different to other ones that I was involved with in that we brought three slightly different disciplines together to make sure we had all angles covered. Within the team we had remastering engineer Simon Gibson working on the restoration work, mixing and recording engineers working on the digital transfers and sort of overseeing the whole thing, and then mastering engineers, including Steve Rooke, completing the final touches. The stakes were high working on such an important project, so we made sure we never had just one individual working alone. There was always at least a couple of people around so discussions could take place on what you were doing and why you were doing it. This included playing what you had done to others, harnessing their feedback, and making sure you had group consensus before proceeding.

The remastering process was fairly uniform across all Beatles' albums and occurred across different rooms and involved several people. My role as an engineer was to oversee the whole process. The first part of the workflow involved the digital transfers from original ¼-inch master tapes. The stock of EMI Tapes from the 1960s has really survived well and were in unbelievably good condition and did not require any baking. However, a few edits needed to be remade where the original editing tape had started to come undone to make sure the tapes played well. The transfer process involved lining up the tape machines to calibrate how they would play back different frequency ranges. As there are no test tones on The Beatles' master tapes to ensure the correct calibration, there was a lot of trial and error. We did a few transfers with different line-ups, and we would compare them. Also, the original BTR tape machines used to record the master tapes were not in great condition, so we ended up using Studer A80s. The choice of analogue to digital convertors was also an important decision during this process. As we were converting from analogue to 24 bit 192kHz sampling rate (the highest available at the time) into Pro Tools, we auditioned a few different convertors and ended up using the PRISM Sound ADA-8XR system for all albums as it sounded the best.

The next stage, after creating the best digital transfers we could, was audio restoration undertaken by Simon Gibson using CEDAR Retouch. The engineer, myself or whoever else was working in that role on that day, would listen through in detail on

FIGURE 2.3 Sam Okell [Photograph], received from Sam Okell

headphones and make notes on everything we thought should be addressed in the restoration including clicks, pops, dropouts and distortions. De-hissing though wasn't really an issue as these are great sounding records and we were just looking to improve on what was already there. Recording with tape can create little clicks and other noises through dropping in and out. Sometimes they are imperceptible, but when you go through and take out a number of these tiny things, the overall difference is quite significant. All these tiny imperceptible things add up, so you really notice the difference. It sounds a bit cleaner. Although it is a brilliant digital audio restoration tool, CEDAR Retouch only worked at 96kHz (at the time), so we ended up chopping up the sections Simon worked on and dropped them back into the 192kHz master to maintain the highest resolution and sampling rate for the project. It was a painstaking process, but we were just trying to do the absolute best we could with the tools available.

Once the restoration was completed, the next stage was to apply analogue EQ as we thought this would work and sound best. I would work with the mastering engineer Steve Rooke, and we would listen to EQ changes we had applied and appraise what we had done with a modern ear. We would question whether making it a bit brighter made it sound better. For example, how did boosting 12–14kHz positively or negatively impact the balance of the instruments, how were the hi-hat and cymbals affected, were the vocals now too sibilant? All these types of decisions were made during this stage.

Another consideration during this stage was the original vinyl format that these recordings were cut to. The cutting engineers would have had to deal with limitations surrounding the amount of bass, requisite time they had on the side of a disc, width and phase. Those considerations would have played a part in how the vinyl originally sounded. Although we issued vinyl versions in 2009, these considerations were no longer valid when we remastered to CD. We therefore had to make decisions outside these previous limitations. For example, would it sound better on 'Taxman' if we boost the bass around 60Hz on the kick drum, making it a bit punchier? The main EQ unit we used was the PRISM Maselec MEA-2 stereo equaliser which is fantastic. We also used some period EMI TG EQ modules which would have been used on the original recordings and they sound fantastic also, but with just a bit less detail than the MEA-2. I don't recall using any digital EQ.

Dynamically, I don't think there was much compression added at all. There may have been a touch of stereo widening but there certainly wasn't a blanket approach. Some of the early albums are just really a two-track multi-track that weren't intended to have the band hard panned left and the vocals hard panned right. That two-track multi-track was intended for a mono release only. I think for these early records we probably actually brought in the panning which we thought sounded more cohesive and the vocals weren't totally disassociated from the band. And there may have been a few times when things have been widened a little. There's a great EMI analogue bit of kit called a Stereo Spreader, from the 1970s I think, which does a bit of mid-side spreading, but it's quite clever in that it doesn't touch the lower frequencies, and as the frequency increases it would increase the spreading. It's a great way to do stereo widening without making thigs sound phased and weird. Remember, it's not until *Abbey Road* that you get the drums across more than one track.

The decision was then taken, in retrospect I'm not sure it was a correct one, that because the primary format then was CD, and the delivery format was 16 bit 44.1kHz, we would send the signal from Pro Tools through our analogue processing and then back into SADIE at 20 bit 44.1kHz so we didn't have to do any further digital conversion for the CD exports.

The final stage was working out the levels through digital limiting. We took the decision to add around 2–3dB of digital limiting to just raise it a little, as 3dB louder does make a difference, particularly when listening on a laptop and comparing something current against a recording from the 1960s. We were just trying to nudge it up into a similar world as modern recordings without going too far and destroying the dynamic range. There are always going to be opinions around this, do you do nothing, or do you make it as loud as a modern record? We decided we didn't want to do that, which I think was a sensible compromise.

We would then make copies for Ringo, Paul, Olivia Harrison and Yoko Ono to listen to everything generally away from the studio. However, we have had situations where we set up listening sessions for them at Abbey Road, in particular 5.1 surround mixes, where they may not have this set up elsewhere. The Beatles is a political situation and if Paul starts saying, 'oh, you know, do this to me' and Yoko starts saying 'John should be this', well I think they're all very aware of that, and thankfully they trust you as part of a professional team experienced to work on it. They have final approval, and I would say on the stuff we've done, 95% the time they've just said, 'Yeah, we think it's great'. Occasionally they have a comment.

It is interesting because I've worked with Paul on his solo records, remastering and remixing, and because it is just him and he has total control, he wants to experiment more in the studio and play around. So, you could say maybe if all of The Beatles were alive today, they would all come into the studio and go 'what can we do with this' and provide more feedback, but I think given the situation, they don't get too involved actively. I think there is a lot of trust with a team appointed by Apple at Abbey Road Studios that they are doing the right thing at the right time with the technology available. Having said that, we did set up sessions where they could compare the original against several demos in short spurts of 15 seconds to hear EQ and level differences, and I recall in general they were genuinely excited. They said things like 'this sounds great, punchier and more polished' and 'people are going to enjoy this'.

I was very happy with the final product as I thought we did the job well. I don't think anything was done that was pushing it too far. I get a lot of feedback on these things and there's always going to be people who say, 'why would you touch it' and 'leave it alone'. That's fine but what is it that they are listening to now? Are they listening to a record that's 60 years old? How great does that sound – particularly if it's been played 10,000 times? Are they listening to the original CD from 1987 that went through some pretty poor analogue to digital converters? So, what are they actually listening to? If we can provide a good translation of the master tape with today's technology, then I think that's the best listening experience you get. And then there are going to be people that can't tell the difference or hear what we have done. In a way that's a good thing because it sounds as they remember it and you haven't changed the essence of it.

The most radical Beatles' project was the Love show where George and Giles Martin mashed stuff up, detuned and time stretched things. Some people loved it because it was different and interesting with new versions of classic Beatles' songs, and others hated it!

Remixing The Beatles (Sam Okell)

I was involved in remixing The Beatles' albums *Sgt Pepper's Lonely Hearts Club Band, White Album, Abbey Road* and *Let It Be* for their respective 50th anniversary releases, and you do go into those discussions of what should we do? But the problem is, if you allow yourself to do one thing, then where do you stop? If I've been mixing a modern record and then I move to remixing a Beatles record, it's very tempting to change things as you would for a modern record, for example, fixing up timing and tuning issues. In the end it was quite a simple decision as we all agreed to not touch the timing or tuning of any Beatles recording. And obviously with remixing there's an awful lot you can do compared to remastering as you have access to the multi-tracks, so it's best not to go down the retuning and timing path otherwise you'll never stop.

You're trying to take a great record and make it sound better. You're not trying to reinvent it or change the intention of what it was, and our decision is always that the intention, the performance, the way they did it, is the essence of the record. For example, 'A Day in the Life' is a great sounding record, but in the middle section where Paul's got his vocal 'woke up, got out of bed' it's kind of muffled sounding and is a little odd by comparison to the rest of the song. And you must take a decision was that done on purpose or should I brighten that up trying to make it sound like clean poppy vocals and match the rest of the song. I think it was an intentional sound they were pursuing at the time of the original production, and in comparison, with John's lead vocal, you must leave that sonic difference between them and respect that.

You can get paralysed with the fear of it all, as a lot of people are going to listen to this and have an opinion on it. Therefore, I believe you need to be quite straightforward and know what you are trying to do, how are you trying to do it, and work within parameters you've set for the project. When I am remixing, sometimes I start off thinking I am going to reinvent this and get it sounding great without listening to the original production. And then when I A/B against the original production I realise I have totally pulled it apart and it's not that song anymore. I then go back towards the original carefully listening to the positioning, EQ, reverb and start matching all that stuff and then I find quite often you swing back, and you end up with exactly the original version. So, you move backwards and forwards between those two places and try and find something that hopefully brings something new to the song?

With an album like *Abbey Road*, there was less to do in a way as it is a proper stereo recording with the vocals centred and the panning like a modern style and it's a good sounding record. With an album like *Sgt. Pepper's Lonely Hearts Club Band*, there's a lot more you can do as they recorded on four-track, filled it up and bounced those four tracks onto one track on another machine and so on. I think they filled four four-track machines by the end of it. So, mixing that album they were limited as they've only got four tracks available to mix with 20–30 different elements across those four tracks, and they can only position those tracks in four different places. They might have put all of the rhythm tracks to the left, vocals on the right and guitar solos in the centre for example. But now we can go back and sync up all those generations of tapes and suddenly we've got an 18–22-track-wide multi-track to mix from. You can do a hell of a lot more in remixing that album now than they could in their original final mix. It's not that you're better than they were or anything like that, it's just now you've got all those things at hand on separate faders that they wouldn't have had.

That 1960s way of working on The Beatles is a great lesson in getting and committing sounds as you go. If the band is all being recorded at the same time to be bounced down to one rhythm track, then the drums have to sound great and be in balance with the bass, and the guitars have to have the right amount of reverb, because that is what it's going to be on the record. I try to apply some of this approach when I am making modern records. Even with a full multi-track at my disposal, I quite often do a drum mix on the auxiliaries through a mono Fairchild compressor and if I think the sound is great, I'll end up using it.

Comparative listening

I asked Sam Okell to undertake comparative listening between a digitised copy of the 1969 vinyl first pressing, the 1987 remastered CD and the 2009 remastered CD for *Abbey Road* for a selection of songs and identify sonic differences they perceived.

Sam Okell

I think there are two main things at play here. For the 2009 remaster, taking out any distortions, clicks and noises that you don't want, I think, provides an overall better listening experience. To me it just sounds a little bit cleaner, and the ear isn't distracted by these things, and this allows you to focus on the music more. The equalisation difference can also be quite dramatic, even in remastering, despite not being able to control the individual elements as you can push the midrange up which brings the vocal up or you make it sound a bit brighter. Overall, I think the 2009 remasters make a better listening experience overall, through mainly the noise reduction and EQ changes.

Digital audio analysis

Having sourced and generated a digitised copy of a 1969 UK vinyl first pressing of *Abbey Road*, I was able to compile 16bit/44.1kHz WAV files for all three versions that were directly comparable and in the same format. The first piece of analysis undertaken was a waveform comparison, as adopted by Barry in his work (2013). I decided to focus on three tracks from Side 1 in their correct order of 'Something', 'Octopus's Garden' and 'I Want You (She's So Heavy)' to determine how the mastering (and remastering) may vary from the outermost track of a vinyl album to the innermost track and to identify any consistencies (or inconsistencies) for a full album side in general.

Figure 2.4 is a screenshot that depicts the loudness/amplitude range over time of the respective waveforms for each release of *Abbey Road* within the Pro Tools digital audio workstation (DAW) software environment. The top horizontal track represents the 1969 vinyl transfer, the middle track depicts the 1987 CD release, and the bottom track portrays the 2009 remastered CD release. As shown, each track consists of the individual songs 'Something', 'Octopus's Garden' and 'I Want You (She's So Heavy)', in that order horizontally and aligned directly above and below each other. The visual representation of the 2009 remastered CD release clearly shows a much louder signal when compared to both the 1969 digital transfer and the 1987 CD version, with the 1987 CD version appearing to be slightly louder than the 1969 digitised version, which is to be expected. Furthermore, the waveform shapes of the 2009 CD version appear more 'block-like' than both the 1969 and 1987 versions suggesting a more compressed signal

FIGURE 2.4 Waveform view of 1969, 1987 and 2009 versions of recordings from *Abbey Road*

overall with slightly less dynamic range. To explore these variances further, I measured left and right true peak level meter readings of the *Abbey Road* 1969 digital vinyl transfer, 1987 remastered CD and the 2009 remastered CD, similar to Barry's analysis previously undertaken on other recordings by The Beatles (2013).

Table 2.1 represents the left and right true peak level values for the three selected songs across the three versions. For the 1969 digitised vinyl version we can see that the true peak level measurements indicate a trend on Side 1 where the album is getting louder (apart from 'Octopus's Garden' right side peak reading) with 'I Want You (She's So Heavy)' approximately 2dB greater in peak value as opposed to Track 2 'Something'. When we view the 1987 CD measurements, we can see that 'Something' and 'Octopus's Garden' are fairly similar (particularly on the right-hand side) with 'I Want You (She's So Heavy)' slightly greater. The 2009 remaster clearly shows the peak levels are much higher, tighter and with less distance between the different tracks, compared to the 1969 digitised vinyl version.

Although this is helpful information in terms of identifying the loudest peaks across an individual track, it does not provide an average peak level for an arguably fairer 'loudness' comparison. To explore these average variances in loudness, I also recorded the left and right root means squared (RMS) level meter readings to display the average level of loudness overall. RMS metering is useful for indicating if two or more songs are approximately the same loudness level (Owsinski, 2008).

Table 2.2 displays the RMS levels for both left and right channels across all three releases. The difference in average loudness between the 1969 digital vinyl transfer and the 1987 CD remaster does not appear significant, with the CD approximately 1–2dB

TABLE 2.1 True Peak level measurements *Abbey Road*

Song	*1969 digitised vinyl*		*1987 remastered CD*		*2009 remastered CD*	
	Left true peak	*Right true peak*	*Left true peak*	*Right true peak*	*Left true peak*	*Right true peak*
'Something'	-3.45	-4.05	-4.21	-3.89	-0.26	-0.28
'Octopus's Garden'	-2.13	-4.80	-4.85	-3.88	-0.29	-0.62
'I Want You (She's So Heavy)'	-1.33	-2.78	-3.75	-3.49	-0.32	-0.30

TABLE 2.2 RMS level measurements *Abbey Road*

Song	*1969 digitised vinyl*		*1987 remastered CD*		*2009 remastered CD*	
	Left RMS	*Right RMS*	*Left RMS*	*Right RMS*	*Left RMS*	*Right RMS*
'Something'	-17.73	-18.09	-16.23	-15.87	-12.03	-12.84
'Octopus's Garden'	-17.32	-19.94	-16.22	-17.12	-11.62	-14.78
'I Want You (She's So Heavy)'	-16.76	-16.58	-15.80	-14.99	-11.79	-10.77

louder on average. However, the 2009 remaster is significantly louder on average, approximately 6dB louder than the 1969 vinyl and 5dB louder than the 1987 CD remaster on 'Octopus's Garden' for example.

The next measurement undertaken was LUFS, a loudness measurement similar to RMS in terms of calculating average loudness, but which takes into consideration human perception of audio loudness and is an interleaved measurement (not separated by left and right).

Table 2.3 represents LUFS measurements across all three versions. It clearly depicts the 1969 digitised vinyl version as less loud across all songs compared to both CD versions. The 2009 remaster is louder than the 1987 CD. The patterns of LUFS loudness readings appears consistent across all versions with 'Something' and 'Octopus's Garden' similar, and 'I Want You (She's So Heavy)' slightly louder.

Table 2.4 depicts decibel measurements of dynamic range (DR) across the three releases of the three song recordings analysed. As displayed, the DR score for the digitised 1969 version, across all three songs, is the highest which implies that the DR on the vinyl version is greater than the DR of both CD remasters. The track 'I Want You (She's So Heavy)' portrayed the greatest difference of DR of 3dB between the vinyl and 2009 remaster. This would suggest that perhaps the greatest amount of compression/limiting may have been applied to the 2009 remaster overall. However, for the track 'Octopus's Garden', the 2009 remaster had a slightly greater DR than the 1987 CD so there are variations between the versions.

To further examine the characteristics of a reduction in DR and increased loudness inherent in the digital replicas, as opposed to the original analogue music artefact, the next measurement undertaken was frequency spectrum analysis. Similar to O'Malley's work, I was keen to examine the frequency spread of the three different *Abbey Road*

TABLE 2.3 LUFS level measurements *Abbey Road*

Song	*1969 digitised vinyl*	*1987 remastered CD*	*2009 remastered CD*
	LUFS	*LUFS*	*LUFS*
'Something'	-18.50	-16.55	-12.98
'Octopus's Garden'	-18.66	-16.51	-12.79
'I Want You (She's So Heavy)'	-17.00	-15.47	-11.51

TABLE 2.4 Dynamic range measured in dB *Abbey Road*

Song	*1969 digitised vinyl*	*1987 remastered CD*	*2009 remastered CD*
	Dynamic range	*Dynamic range*	*Dynamic range*
'Something'	11	9	9
'Octopus's Garden'	12	10	11
'I Want You (She's So Heavy)'	12	10	9

versions to identify where these dynamic differences existed and why they were made (2015). It is important to note that the images below only provide a brief snapshot in time on the various frequency and volume levels across all three releases.

Figure 2.5 represents the frequency spectrum for the song 'Something', the second track on Side 1 of the album. The image was captured at around 38 seconds into the song at the start of the second verse where George sings 'Some'. The light grey colour represents the 1969 digitised version, the 1987 remaster is medium grey and the 2009 remaster is dark grey.

It is evident that the 1969 vinyl transfer is the least loud, with the 1987 CD slightly louder and the 2009 remaster significantly louder. All three versions are fairly similar in shape except for the range 5–9kHz where both CD versions, in particular the 2009 remaster, appears to have been boosted. Furthermore, this boost for the 2009 CD appears to be maintained from 9kHz onwards whereas the 1987 remaster shows a reduction, falling to the same level of the 1969 vinyl. Apart from the boost in the 1987 version for the 5–9kHz range, the almost identical shape and levels between the 1969 and 1987 versions appears consistent with the perception that there was not a lot of difference between these versions.

Figure 2.6 represents the frequency spectrum for the song 'Octopus's Garden', the fifth track on Side 1 of the album. The image was captured at around 42 seconds where Ringo sings 'I'd' at the beginning of the first chorus. The light grey colour represents the 1969 digitised version, the 1987 remaster is medium grey and the 2009 remaster is dark grey.

It is evident that the 1969 vinyl transfer is less loud than the CD remasters across the entire spectrum, except for 20–35Hz, where it is louder than the 1987 version. Similar to the analysis image for 'Something', there appears to be a boost in the 5–9kHz frequency range on both CD versions. This appears consistent with the notion that

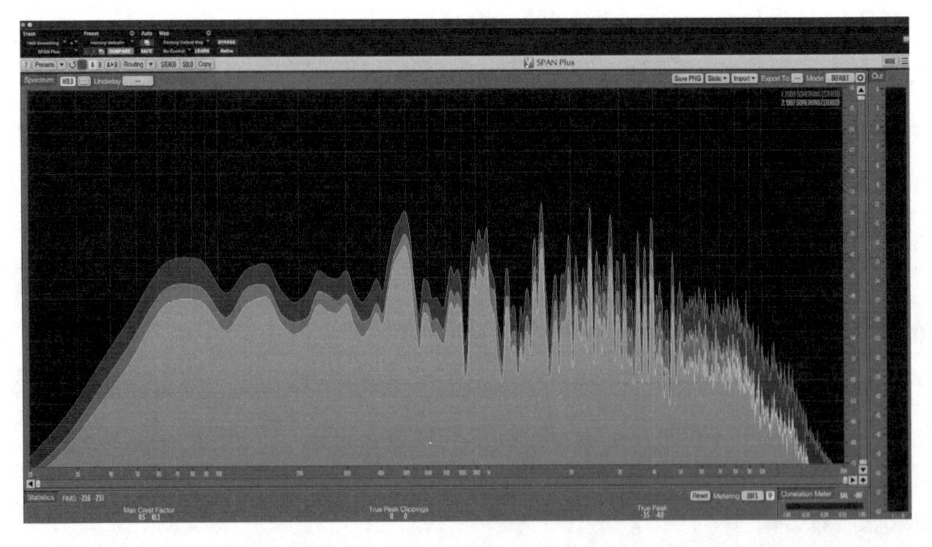

FIGURE 2.5 Spectrum analysis image for 'Something' – 1969 (light grey), 1987 (medium grey) and 2009 (dark grey)

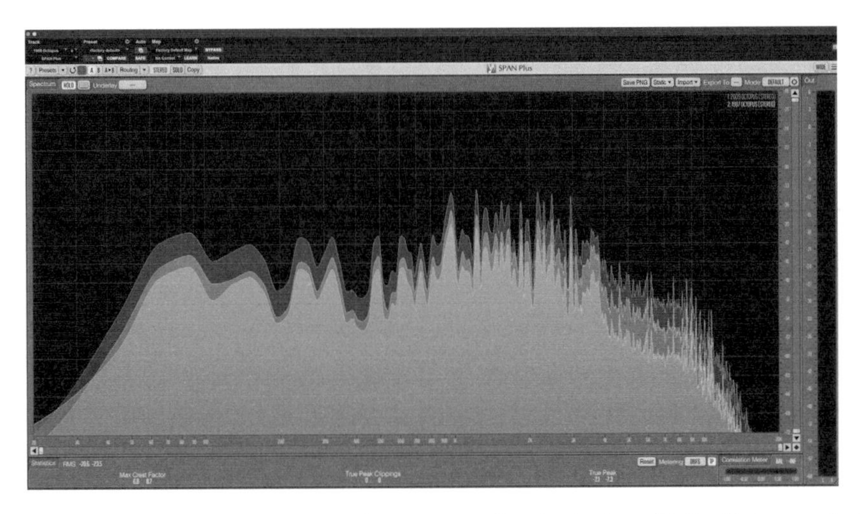

FIGURE 2.6 Spectrum analysis image for 'Octopus's Garden' – 1969 (light grey), 1987 (medium grey) and 2009 (dark grey)

generationally remasters appear to be getting brighter – the 1987 version was brighter than the 1969 original release, and the 2009 remaster was brighter than the 1987 CD remaster.

Figure 2.7 represents the frequency spectrum for the song 'I Want You (She's So Heavy)', the sixth and final track on Side 1 of the album. The image was captured at around 7 minutes and 10 seconds towards the end of the track when all the guitar parts and other instruments are in full swing. The light grey colour represents the 1969 digitised version, the 1987 remaster is medium grey and the 2009 remaster is dark grey.

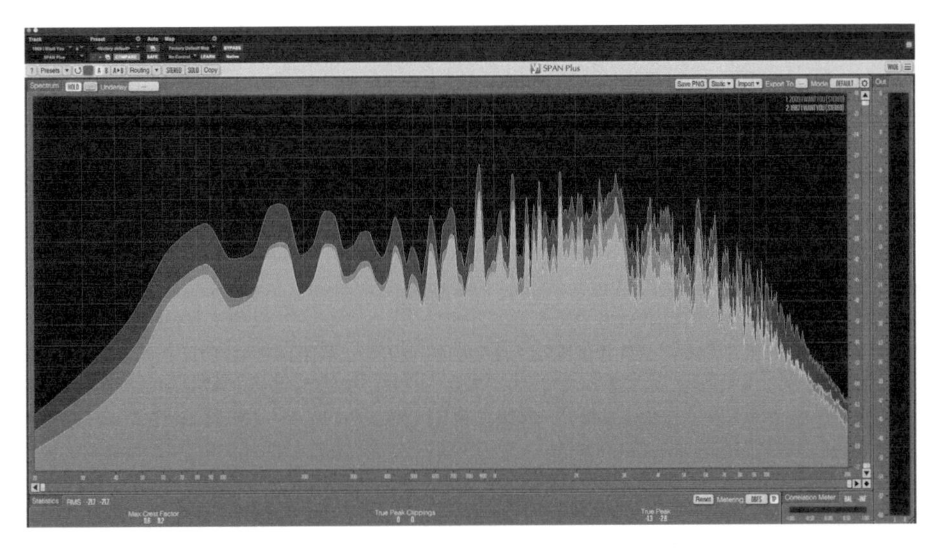

FIGURE 2.7 Spectrum analysis image for 'I Want You (She's So Heavy)' – 1969 (light grey), 1987 (medium grey) and 2009 (dark grey)

It is evident that the 1969 vinyl transfer is fairly similar to the 1987 CD, particularly across the 200Hz–1.75kHz frequency range, with the 1987 remaster slightly louder across the remaining frequencies, with a boost in the 2–6kHz range. The 2009 remaster is significantly louder across all frequencies compared to the 1969 version, with even more emphasis placed in the 20–400Hz and 8–16kHz ranges. This appears consistent with Okell's summation that the EQ differences can be quite dramatic when remastering and pushing up different ranges can make it sound brighter as well as fuller at the bottom end.

Cultural heritage

Just because we can remaster iconic recordings does it mean we should? Is this acceptable from a cultural heritage leaning, and what does this even mean? I was keen to determine how those music production practitioners involved in the original production and/or the digital remaster or replica felt about the original master tapes being 'manipulated' digitally. I was also interested on their thoughts surrounding future generations potentially viewing remastered editions as the original music artefact.

Sam Okel

I think it's OK to remaster The Beatles, but it must be approached sensitively and with a very clear view of what you're trying to do and why. The digital tools are amazing when used in the right way, but you don't want to go too far with these things.

Between us as a team on this project, I think the debate was always, was that sound intended to be there and would they have wanted to remove it at the time if they could? For example, Ringo having a squeaky kick drum pedal. I don't think the engineer or the producer or the band at the time were going 'great we love the sound of that squeaky kick drum pedal', that's really making that track something special. I think it may have been a case of 'fucking oil that pedal Ringo' or something similar. So, I think the squeaky kick drum pedal, everyone would have agreed, could go. I know this is not *Abbey Road*, but I think the one of the biggest debates I can remember was at the end of 'A Day in the Life' on that long piano chord. There is a chair that squeaks right at the end, and I don't recall what the final decision was, but it was roughly half the team wanted to keep it in and the other half wanted to remove it.

I think there is no right or wrong with some of these things, as someone can argue that is what you hear and it is important to the character of the recording, so why take it out. I think most things we took out were little digital clicks and distortions or maybe a pop on a plosive on a vocal, so I don't think there was much debate around those things. Personally, I think the three or four tracks that were recorded on the rooftop for *Let It Be* sound much better without the wind noise and rumble. That album was probably the least technically good sounding album so that probably had the most restoration in terms of trying to take out all those defects in the recording. Furthermore, you have recordings from Twickenham Film Studio, and then they went to Apple Studios in Saville Row where they had a different microphone collection, and some of the large diaphragm condensers that were used did not have pop shields on, so compared to the other albums, this required the most work.

I would question if the remasters would eventually replace the originals for future generations. With something like The Beatles, probably in another 20 years' time there'll be another remastering project where they question is there something more we can do? Are the tapes still in good condition? The Beatles' archive thankfully is well maintained and the original tapes are all still in great condition. The problem you have with other slightly less well managed catalogues of bands is that something may get remastered, and the original media gets lost, and then that version gets remastered and then you start to get a slightly bastardised version of the original records. I don't see the 2009 remasters as bastardised versions of the records; I think they're just slightly polished up for the modern ear. Hopefully if somebody in 30 years' time is listening to the remaster and then rediscovers the 80-year-old vinyl, they won't think the original vinyl is so much better. Everything that's good about that original is on the remaster.

John Kurlander – tape op Abbey Road

I think it is OK for *Abbey Road* to be remastered. The recording of *Abbey Road* was done in 1969 with the best equipment and personnel available at the time. Also, with the presence and participation of The Beatles themselves ... so, a collaborative effort. Any subsequent remaster and release must be considered 'after the fact' and therefore is subject to different influences. Even if the same team were to remix the album again one year later, say 1970, the result could be notably different. As an active musical professional in 2022, I have no interest in re-visiting my previous work and prefer to focus on productions of new music on new equipment.

Furthermore, at this time we have many catalogue works being remixed and remastered for Dolby Atmos which is the current format. However, in a few years I feel this will surely pass and newer release formats will emerge ... maybe more complex or maybe simpler. We see already that Atmos which was invented for large-scale theatrical venues, has been stripped down to a simple binaural format so that the music listening public can enjoy on earbuds.

Regarding future generations potentially viewing remasters of *Abbey Road* as the 'authentic' music artefact, I think this is fine. I cannot predict what the 100-year centennial anniversary edition of *Abbey Road* will sound like when it is released in 2069, possibly a broadcast directly into the brain without the need for speakers or headphones ... maybe in mono. Each generation has a released version for the public and equipment of that era ... and it is all OK.

If you really want to experience the 1969 version in the best and most authentic way, I'm really sorry ... I guess you had to be there!

Bibliography

Barry, B. (2013). *(Re)releasing The Beatles.* Paper presented at the Audio Engineering Society Convention 135, New York, USA.

Beatles Interview Database (n.d.). *The Beatles Ultimate Experience*, accessed 7 April 2022, http://www.beatlesinterviews.org/dba11road.html.

Belam, M. (2007, 17 April). It was twenty years ago today … The Beatles CD reissues from 1987, accessed 7 May 2022, http://www.currybet.net/cbet_blog/2007/04/it-was-twenty-years-ago-todayt.php.

Bieger, H. (2012, November). Abbey Road Studios, London. *Sound on Sound*, accessed 6 April 2022, https://www.soundonsound.com/music-business/abbey-road-studios-london.

Bosso, J. (2022). Beatles engineer Geoff Emerick's Abbey Road track-by-track interview: 'For the first time, John and Paul knew that George had risen to their level', *MusicRadar*, accessed 7 April 2022, https://www.musicradar.com/news/abbey-road-beatles-track-by-track.

Discogs (2022). *The Beatles – Abbey Road*, accessed 8 April 2022,https://www.discogs.com/master/24047-The-Beatles-Abbey-Road

Doggett, P. (1987, June) EMI and The Beatles in 1987, *Record Collector*, No. 94, 11–16.

Golsen, T. (2021, September) Track by track: Breaking down *Abbey Road* in the words of The Beatles. *Far Out*, accessed 6 April 2022, https://faroutmagazine.co.uk/beatles-abbey-road-in-the-words-of-lennon-mccartney-harrison-starr/.

Handley, J. (2019, September) Recording *Abbey Road*: The Beatles' first (and last) album of the modern era. *Reverb*, accessed 6 April 2022, https://reverb.com/uk/news/the-making-of-the-beatles-abbey-road.

Inglis, S. (2009, October). Remastering The Beatles, *Sound on Sound*, accessed 6 April 2022, https://www.soundonsound.com/techniques/remastering-beatles.

Kehew, B., & Ryan, K. (2006) *Recording The Beatles: The Studio Equipment and Techniques Used to Create Their Classic Albums.* Curvebender Publishing.

Lewisohn, M. (1988). *The Complete Beatles Recording Sessions.* The Hamlyn Publishing Group.

Marinucci, S. (2019, September). Beatles engineer John Kurlander remembers the making of *Abbey Road. Variety*, accessed 7 April 2022, https://variety.com/2019/music/news/beatles-abbey-road-engineer-john-kurlander-remembers-1203351236/.

O'Malley, M. (2015). The definitive edition (digitally remastered). *Journal on the Art of Record Production* (10).

Owsinski, B. (2008). *The Mastering Engineer's Handbook: The Audio Mastering Handbook.* Cengage Learning.

Pacuta, W. (2020, 1 February). *Abbey Road* 50th anniversary deluxe edition. *High Fidelity, No* 189, accessed 7 May 2022, https://www.highfidelity.pl/@kts-996&lang=en.

The 1987 CD mixes (2009, 11 September). The Daily Beatle, accessed 7 May 2022, http://webgrafikk.com/blog/uncategorized/1987-cd-mixes/.

Womack, K. (2019). *Solid State: The Story of Abbey Road and the End of The Beatles.* Cornell University Press.

Zaleski, A. (2017, 26 February). The day The Beatles' first CDs arrived in stores. *Ultimate Classic Rock*, accessed 7 May 2022, https://ultimateclassicrock.com/beatles-first-compact-discs/.

3

REMASTERING ELTON JOHN'S *GOODBYE YELLOW BRICK ROAD*

Introduction

The double album *Goodbye Yellow Brick Road* by Elton John was originally produced and released on vinyl in 1973. Having secured seven Top 40 singles in the previous two years, Elton John was riding a wave of success culminating in the creation of this iconic musical artefact, his most commercially successful release of his 50-year career. This album spent eight weeks at the top of the album charts and sold over 30 million copies (Greene, 2014). Although originally released on vinyl, this double album has since been released on various formats including eight-track cartridge, cassette tape, CD, Blu-ray audio and MP3 (Discogs, 2021). This chapter will primarily explore and analyse the production techniques used to create and master the original 1973 release as well as digital remastering practice applied to this iconic recording in 1995. Additionally, sonic comparisons between the 1973, 1995 and 2014 releases will also be investigated through comparative listening and digital audio analysis.

Recording at Château d'Hérouville Studios (near Paris) in 1973

The recording for the album was originally to take place at Dynamic Sound Studios in Kingston, Jamaica, based upon advice from drummer Charlie Watts from The Rolling Stones who had recorded tracks for their album *Goats Head Soup* there. Upon arrival in Jamaica, Elton John, lyricist Bernie Taupin, guitarist Davey Johnstone, bass player Dee Murray, drummer Nigel Olsson, producer Gus Dudgeon (David Bowie, Elton John) and engineer Ken Scott (The Beatles, Pink Floyd and David Bowie) were confronted with a violent atmosphere. As Taupin commented, 'There was barbed wire around the studio, guys with machine guns, people yelling obscenities at us in the street – there wasn't one positive vibe in the place' (Buskin, 2011). Furthermore, the studio did not quite live up to expectations.

DOI: 10.4324/9781003177760-4

> We realised when we walked in the studio that there was no way we could make a proper Elton John record there because it was like two microphones. We did a version of 'Saturday Night's Alright' that sounded like a bunch of angry bees. It just sounded terrible. Ken Scott is a brilliant engineer and if he and Gus couldn't get something happening, it was obvious there was nothing we could do.
>
> *(Johnstone, 2013)*

The group left Jamaica and relocated to Strawberry Studios, housed within the 18th-century Château d'Hérouville located in a village about 30 miles outside Paris. This was a familiar setting for the group with Elton having recorded his previous two albums there with the same band line-up and Dudgeon producing. Elton was obviously impressed with Dudgeon's abilities in the studio to ensure a great sound and performance was captured evidenced in his quote below:

> Gus Dudgeon, I have to say was the fifth member of the band. The actual sound quality of this recording is extraordinary and that was down to him. He was our fifth member, like The Beatles had George Martin, we had Gus.
>
> *(John, 2014)*

Located on the first floor of the Château, the studio consisted of a small control room with a more spacious live recording room. The control room was equipped with an MCI JH 416 console and a 16-track Ampex tape machine and the live space consisted of one room (no isolation booths) with acoustically treated screens to help control bleed from instruments (Buskin, 2011).

The typical studio set-up looking through from the control room window consisted of Nigel Olsson's drums in the far left-hand corner miked up with an AKG D12 on the kick (front skin off), Neumann KM 56 small condenser on the snare, AKG C28 on each tom drum, either Neumann U67 or AKG C12 condenser pair for overheads and whatever directional microphone they had on the hi-hats. Elton John's piano was to the right of the drumkit with a pair of Neumann U67 on the piano and one Neumann U67 for vocals. Additionally, a wooden box was placed over the piano instead of its original lid to achieve greater separation from the drums and to house the two microphones. Dee Murry was set up near left side with a DI on the bass as well as a U67 on the Ampeg bass cabinet. Davey Johnstone was set up near right side with an AKG C12 set up on the guitar amplifier (Buskin, 2011).

The songwriting and recording sessions took place at a fast pace. Elton described the songwriting and recording process in an interview with *Rolling Stone Magazine.*

> During a typical day the band would come down, there'd be instruments around the breakfast table, Bernie would be writing at the typewriter, I'd be sitting at the electric piano, and as the band came down for breakfast, I would write the song, they would pick up their instruments and play it. We'd record about three or four tracks a day. They were mostly made up on the day they were recorded. We were a very tight band with a lot of touring experience. We'd capture more songs in two or three takes. The whole record took about eighteen days.
>
> *(Greene, 2014)*

The completed recordings from Château d'Hérouville consisted of all of the backing tracks, lead and backing vocals, guitar overdubs and any obvious keyboard overdubs. Due to the 16-track limit, tracks needed to be bounced down to accommodate all of the recording materials, in particular drums and backing vocals. A typical bounce-down contained the kick drum on its own track, the rest of the drums bounced down to two tracks, the piano as a stereo pair (two tracks), bass guitar on one track and the guitars on four tracks. There would also be two tracks of backing vocals and one of lead vocals. This left room for overdubbing of orchestral and synthesiser parts to be recorded later at Trident Studios (Buskin, 2011).

Ken Scott was unavailable for these sessions, so engineer David Hentschel was brought into the team. David knew Elton and Gus through his previous session synthesiser playing on Elton John's song 'Rocket Man' from his 1972 album *Honkey Château*, as well as through engineering bands that Gus had produced.

David Hentschel on recording *Goodbye Yellow Brick Road*

Background (David Hentschel)

I began my career at Trident Studios as a tea boy as a 17-year-old in my gap year between finishing school and starting university. I studied classical music at school and got A-levels in science and maths and was keen to undertake the course at the University of Surrey (the only one offering such a degree) that combined music and technology. Well, on my first day at Trident I was making tea for David Bowie who was recording *Space Oddity*, with Gus Dudgeon producing, and I thought they didn't tell me about these types of jobs in the school career office. Needless to say, I fell in love with it and stayed, ditching university after about six months having already been promoted to assistant engineer by this stage. I learned my craft through watching other people and asking plenty of questions. I was exposed to a number of great engineers and producers across a wide range of musical styles and therefore picked up a lot of knowledge around appropriate microphone choices, techniques and placement.

Early on the studio manager figured out that I had a musical background and would let me use the studio on my own during downtime to record friends from my school band. We would wait for Queen or some other artist to finish recording at say 3:00am and go in and record until I had to be back at work again at 8:00am. They also let me take four-track and eight-track tapes out of the tape library and practise mixing in the mix room on my own – again in studio downtime. Sometimes during these sessions other engineers who were around at the time, Ken Scott (The Beatles) or Roy Thomas Baker (Queen) for example, would pop their head in see what I was doing, and I would ask them what they thought of my mix and they would say things like 'it sounds great, maybe just try doing this or adding that' which was a great way to learn.

I also learnt how to communicate professionally and appropriately in given situations. I learnt when to say the right thing and when to say nothing at all which was critical for ensuring sessions ran smoothly and great performances were achieved.

Recording at the Château d'Hérouville (David Hentschel)

On a technical level the studio was a fairly basic 16-track studio, having previously been the personal studio of French composer Michel Magne, with a small selection of good quality microphones. However, it was a lovely spot and an inspirational set-up. From the control room you looked through into the studio, and then there were windows all around the studio looking out over the French countryside. As we were living there all of the wives and girlfriends came over and stayed with us, there was a pool, the catering was great – it really was a great way to spend three weeks, like a big happy family really.

Working on this album with Elton was easy as he was the complete professional. He knows what he wants very quickly from writing through to the recording and we would therefore get the songs recorded in two or three takes. We would then do a couple of basic overdubs, then record his vocal after which he would go off into Paris shopping. It was great. I mean we just got on with it. He was so fast and so together, you know.

In many cases the band would be having breakfast while Elton was writing on the piano close by, so they had probably heard it 20–30 times by the time we got into the studio shortly after. They kind of all automatically fell into the right parts. Nigel tended to play pretty straight, and Dee was just a great bass player and used to come up with these wonderful melodic lines while still laying a credible foundation. Davey was very adventurous on the guitar and Elton played his part. There was no click track

FIGURE 3.1 Elton John and David Hentschel recording at the Château d'Hérouville [Photograph], courtesy of David Hentschel

in those days so Nigel, as a great drummer, knew not to rush his fills and Dee would sit just behind the beat as all good rock bass players do.

Elton was the ultimate professional when it came to recording his vocals. Like most things these were done quickly, with no comping; he just got on and delivered incredible performances that were full of expression and dynamics. They were a very professional and tight outfit who communicated well with each other and with us during those sessions and I believe this is reflected in the quality of the performances and recordings. We'd be working 10–12 hours per day normally, finishing up at around 9 or 10 o'clock, when I'd occasionally have a hit of table tennis with Elton. It was a very happy time conducive to evoking great song compositions and performances.

I can still remember watching Elton compose the music for 'Candle in the Wind' on the piano while I was finishing my breakfast and knew that it was, amongst other songs on the album, destined to be a shining light, but by no means the only one.

Recording at Trident Studios, London in 1973 (David Hentschel)

Although the majority of recording had been completed at the Château d'Hérouville, Hentschel and Dudgeon returned with the multi-track tapes to Trident Studios in London where they teamed up with assistant engineer Peter Kelsey to complete all of the recording and mix the album.

> Once the basic tracks, guitar overdubs, and lead and background vocals had been recorded at the Château, it was up to Gus and David to put on the finishing touches and mix the album. This took place at Trident Studios in London (the site of Elton's album sessions prior to France). Percussionist Ray Cooper was brought in to play tambourine on 'All The Girls Love Alice' and arranger Del Newman (who had worked with Elton when he was still Reg Dwight on some demo sessions in 1968) did orchestral arrangements on six of the songs.
>
> *(Eltonjohn.com, 2013)*

The recording was completed in the live room using the custom-made Trident A-range console with a 16-track 3M tape machine (Buskin, 2011). Additional sounds including the horse galloping by on 'Roy Rogers' were recorded on location using a portable Nagra tape recorder (David Hentschel, personal communication, 2021).

Recording 'Funeral for a Friend' (David Hentschel)

As we were recording at the Château d'Hérouville, it became evident through Bernie Taupin's lyrics that strong cinematic themes were developing including references to Marilyn Monroe and Roy Rogers. The idea at one point was to call the album *Silent Movies and Talking Pictures*. From there, it was decided that it would be a good idea to include a synthesised version of the motifs associated with either 20th Century Fox or Columbia Pictures, but they were unable to secure the clearances from the movie production companies to use their music themes. Gus then came up with the idea for me to create a short overture to the whole album, that is to arrange some melodies from some of the songs into a synthesiser introduction.

I sat at home and played around with the structure of a few songs, just little snippets of the melodies from 'The Ballad of Danny Bailey (1909–1934)', 'I've Seen That Movie Too', 'Roy Rogers' and a couple of others, I think. It was sort of like doing a crossword, similar to the way David Bowie used to sometimes compose lyrics by writing words and little phrases on pieces of paper and shuffling these around. After I was fairly happy with what I had, I started thinking about the tonality of the different sections, as I wanted to make it as moody as possible. Those big fat sounds I referred to as 'strings' for lack of a better word, but I mean it didn't sound like strings at all. It sounded like swarming bees! That's how I wrote it out, the monophonic melody with interlocked harmony and references to tonality.

I used the ARP 2500 synthesiser (a competitor to the more popular Moog synthesisers of the time) for recording. I used to sit at this thing for hours just fiddling about, learning new things, trying to create new sounds through modulation, playing with arpeggiation – it was just a wonderful toy to have. At the time we were limited by the amount of processing tools and boxes we had in the studio, for example we could record Leslie rotating speakers and/or abuse tape machines to do phasing and flanging, but there wasn't a lot else available to us. With the ARP however, you could think of a sound, something that has only existed in your imagination, and using your ears you could try to mould and sculpt this sound. It was wonderful. I was the first person to use one in the UK.

'Funeral for a Friend' was recorded in one day in Trident Studios 'live' room on a 16-track tape machine with me playing one note at a time. On that particular synthesiser, you could play up to three notes, but it was a very early analogue model and it kept going out of tune and was difficult to retune. So, I decided it would be easier for me to write out all of the parts, and then play all of these parts monophonically, as if I was a player in an orchestra. This also allowed me to use one hand to play the notes and the other to adjust the filters and the amplifiers which was what you needed to do back then to introduce dynamics. All of the sounds you hear in this short musical piece were created and played by me on the ARP. The wind sound is pink and white noise with a filter adding a certain amount of resonance and the bell sound, I recall, was probably a ring modulator.

Mixing *Goodbye Yellow Brick Road*

The album was mixed in the Mix Room at Trident Studios. The room was upstairs and included a 20-channel Sound Techniques console, a Studer 2-inch 16-track tape machine, a Studer ¼-inch two-track tape machine, large CADAC monitors, UREI LA2A compressors, a UREI 1176 limiter and EMT reverb plates.

Peter Kelsey on mixing *Goodbye Yellow Brick Road*

Background (Peter Kelsey)

I shared a flat with Roger Taylor, the drummer from Queen, who were recording their debut album at Trident Studios. I was friends with the band and paid them a visit at the studio and chatted to their engineer and asked about getting a job. I filled in an application and got a job as a tea boy. It was a tough job as you started at 9.30am and finished when the last session finished, which could be 9.30am the next morning when the next guy came in, so yes, 24 hours on and 24 hours off basically. I would make tea and coffee for everybody but also help the tape op and/or 2nd engineer. I would help prepare sessions and pull them down at the end, so I got to learn what microphones they were using on what, and where they were placing them. After about six months I was promoted to tape op for Carly Simon's album *No Secrets*. During those recording sessions I asked the engineer Robin Geoffrey Cable lots of questions and watched, and that's how I learnt. From there, around six months later, I was promoted to become an assistant engineer and assigned to work on *Goodbye Yellow Brick Road*.

Mixing Goodbye Yellow Brick Road *album (Peter Kelsey)*

I like to think of mixing as a performance, particularly mixes of this vintage. It is as if the musicians have all finished their performances and now it is our turn. You have to remember that when we mixed *Goodbye Yellow Brick Road*, and other records around that time, that there was no automation. Therefore, changes during the mix, gain, effects, panning etc had to be made 'live', as you were in the process of recording your mix to the ¼-inch tape.

For *Goodbye Yellow Brick Road* album, Gus would assign us different sections and/or faders of the console. Generally, Gus would take the drums and bass, I would be given either guitars or vocal and David would do the rest. We had to get to know each song really well and work as a team, perform as a coordinated unit like I mentioned before, to achieve the sonic changes and final mix that Gus was searching for. For example, Gus liked to make the drum fills stand out so every time there was a drum fill, he liked to push those faders up and then drop them back down to where they were. As we were monitoring other areas of the song with various changes required, we all had to know the song when the drums were going to rise and fall, so we could adjust other sections so as not to interfere with or detract from the prominent parts. We would record it, wait a couple of minutes and then play it back to see if we got the exact mix Gus was after. If we didn't, we would do another take.

It's like I said before, it's a performance. You have three people performing together to get the desired sound. I prefer this way of mixing because it's fun, it's creative and you feel like you perform something you have done.

I met Elton John a couple of times when he visited the studio, but he wasn't really involved at the mixing stage. All decisions on mixing came from Gus.

Specific tracks

'Funeral for a Friend/Love Lies Bleeding' (Peter Kelsey)

After I had recorded all of the synthesiser parts for 'Funeral for a Friend', we had to mix the end of that to lead into 'Love Lies Bleeding'. I'm not sure if people can remember but there is a castanet sound and because we didn't have any tracks left on the tape, that was played and recorded live during the mix. Once David had created the sound, he went over to the synthesiser to play one specific note to trigger the castanet sound which was captured live on the fly onto the ¼-inch tape.

'Goodbye Yellow Brick Road' (David Hentschel)

I had to remix this track based upon a guitar effect Davey Johnstone had used. It had a Lesley rotating speaker effect and I think it may have been called a Mu-Tron? Anyway, it was the last song on the first side of the first record, and because it was near the centre of the disc and there were a lot of transients in that guitar sound, the mastering engineer couldn't cut it as it was distorting to vinyl. So, I had to go back and recompress those elements and change the stereo perspective to have it mastered to vinyl. But that's one of the only effects I think Davey really used to be honest.

'Benny and the Jets' (David Hentschel)

All of the noise and live sound on that track was fabricated and we did that in the mix. Gus got hold of some applause from a Jimi Hendrix concert in the Isle of Wight. We had the roar of the audience coming out and the sort of murmuring at the background and had a stereo feed of that we pushed into the mix. The clapping in time was Gus, Peter and me. We deliberately clapped slightly out of time, cut it all up, double and then quadruple tracked it, and then sent it through a digital delay line, a new invention at the time, to get that slapback effect. All of the whooping, hollering and whistling was Gus in the background, and we also recorded takes of stomping our feet on plywood. So yes, it was all completely faked.

Mastering *Goodbye Yellow Brick Road*

The album was mastered, or 'cut' as the process was referred to then, by Ray Staff at Trident Studios in their mastering suite.

Ray Staff on mastering *Goodbye Yellow Brick Road*

Background (Ray Staff)

I started off as what they called a 'junior runner' at Trident Studios. After about a year I was offered the opportunity to work in the tape copying room and to assist in the disc cutting room. In the newly formed Mastering Department, I gained experience by assisting the mastering engineer Bob Hill to cut records. Eventually, after gaining his trust, I was given the opportunity to work on my own. I was also given more technical training by Sean Davies at a later date and became Trident's first Chief Mastering Engineer. I cut records at the time for artists including David Bowie's *Ziggy Stardust and the Spiders from Mars*, Elton John's *Goodbye Yellow Brick Road*, The Rolling Stones' *It's Only Rock 'N Roll*, Led Zeppelin's *Physical Graffiti* and The Clash's *London Calling*. Mastering was evolving fast at this time and knowledge and technique was often gained by experience and listening to the work of my contemporaries.

Mastering to me is the ability to transfer a recording into a delivery format ready for the manufacturing process with the quality of reproduction acceptable to the producer and the artist. This may require the modification of the sound by means of equalisation, compression etc. If they are content with the sound of their final masters, the challenge is to produce a master that will reproduce as close as possible the same quality of sound as the original. This can be challenging when producing a master on a different format from the original recording.

Trident Studios equipment (Ray Staff)

The equipment used at Trident Studios for cutting and mastering included a Neumann VMS66 with VG66 valve disc cutting amplifiers and an Ortofon STL 631 valve high frequency limiter. There was a customised control desk made which included and housed Neumann equalisation modules, a Neumann elliptical equaliser (to reduce vertical low frequency signals), a Fairchild 670 compressor, a Studer C37 ¼-inch tape recorder a Dolby A301 noise reduction processor and Tannoy 15-inch red speakers in Lockwood cabinets powered by Quad II valve amplifiers.

Mastering Goodbye Yellow Brick Road *(Ray Staff)*

Unfortunately, this took place a long time ago and the details are lost to time. Cutting notes were often held at the studio, and when Trident closed, the notes were most likely lost. Having said that, the general workflow we followed at the time consisted of the Studer tape machine feeding into the Dolby noise reduction system, then through the Fairchild compressor and then onto the Neumann EQ modules and elliptical EQ to set levels/EQ. From there, the signal was sent through the Ortofon high frequency limiter for adjustment and finally arriving at the Neumann cutter.

Each song would then be listened to and adjusted if necessary. The first requirement is for the level from one song to remain consistent and be appropriate for the next one. For example, a ballad might be slightly lower in level. If a track lacked bass, treble or mid frequencies, we might compensate for that with EQ adjustments. The level would also need to be optimised for the appropriate cutting level which would be determined by duration, format (7-inch or LP), stereo width and/or bass content. A queue list would then be made, and adjustments were made on the fly as we cut in real time to vinyl. With regards to gaps and fades between songs on an album, some producers would prefer a set duration of say two or three seconds, but most were determined by the musical flow and timing. Generally, we would normally cut a vinyl double album during the course of a normal working day.

I do recall though for *Goodbye Yellow Brick Road* that Gus generally wanted the vinyl to sound as close to the original recording as possible. If we added any equalisation, it would have been minimal. I seriously doubt we would have used any compression. At the time we had a very limited range of equipment available to make any substantial changes. For example, the incremental steps were 2 or 3dB and the selectable frequencies were minimal when compared to more modern analogue equalisers which had multiple selectable frequencies, Q and incremental steps in the order of 0.5dB or less. Although we had a beautiful Fairchild compressor, it was not ideal for all recordings, and modern rooms often have three or more to choose from.

As far as I remember everyone was happy with the vinyl pressings. The Studer valve tape machine was identical to the machine the album was mixed on and the Neumann valve amplifiers had a lovely warm sound. It was a nice system. I am happy with the quality of what we produced given the instructions and guidance we were given by Gus at the time. At a future date he might have revised his opinion and wanted to enhance the reproduction using newer and more versatile equipment and/or alter the sound to suit new digital formats.

Thoughts on remastering (Ray Staff)

I have remastered many original recordings myself. In some instances, I have remastered projects I originally mastered in the 1970s some 40+ years later! In most cases I was content and happy with the work we originally did although with some I was slightly disappointed. However, any remastering will always sound different unless you have identical equipment to that used in the original mastering which is an unlikely situation. Another consideration with remastering which needs to be considered is the possible deterioration of the original recording.

The producer for the remastering might well want to change the sound. It is not uncommon for the original producer not to be involved and the guidance to be totally different. I have also been very surprised by what a producer might want to change and how their opinion of the recording has changed with the passing of time. They can sometimes express how disappointed they were with the original mix and want to achieve something substantially different from the original mastering.

This can be extremely time consuming, and I tend not to be critical of remastering carried out by other engineers given the variable issues they have to contend with. In the case of remastering *Goodbye Yellow Brick Road*, I would respect any changes made if it was supervised as approved by Gus.

Remastering *Goodbye Yellow Brick Road* for CD in 1995

Goodbye Yellow Brick Road was remastered for CD in 1995 as part of the *Classic Years* series where other Elton John albums were also remastered at this time. The remastering was undertaken by Tony Cousins at Metropolis Mastering and produced by Gus Dudgeon.

Tony Cousins on remastering *Goodbye Yellow Brick Road*

Background (Tony Cousins)

I got into studio work through playing in bands and getting to know people. This was around the late 1970s and I started as a tape op copying tapes, editing tapes, lining up tape machines and preparing the tapes for various formats. For example, if a song was going to be played on the radio locally in the UK and/or Europe, I would change the EQ on the machine for the International Electrotechnical Commission (IEC) standard to make a copy onto a 7½-inch tape spool which would be sent to the radio station. I would also send tapes to the cutting room where they would be cut (mastered) and then they would come back. I was able to listen to what had happened, the differences between the original mixed version and the cut version I got back, and that's how I basically became interested in this area. I was really keen to find out how they appeared to even out all of the elements on the tape. This was at Townhouse Studio, which was Virgin's flagship studio at the time, and the cutting room was very successful, the main one around. It was run by a guy called Ian Cooper, who has since retired, but was one of the best cutting engineers in England at the time. They opened up another cutting room and I said I would like the job.

In 1993, I started up Metropolis Mastering with Tim Young and Ian Cooper. The team have mastered many recordings and artists including Oasis, Ed Sheeran, Led Zeppelin, Arctic Monkeys, George Michael, The Who and Pet Shop Boys. Personally, I have mastered Adele, Robbie Williams, Peter Gabriel and of course, Elton John to name a few.

Thoughts on remastering (Tony Cousins)

When I started cutting, I became exposed to all sorts of issues. For example, some of the tapes were in worse condition than others which involved extra work in trying to get the tape to run smoothly. Furthermore, one track may need to be raised 2dB and the next track may have to be lowered to get them sounding consistent against each other. Mastering taught you to match up tracks against each other not only level wise, but also frequency wise as well as how to transfer this onto a vinyl disc.

It's interesting this current vinyl resurgence because I fear some of the modern engineers may be mastering to vinyl for the first time and/or are unaware of limitations of the format. For example, the need to use elliptical equalisers to control the movement of the groove, out of phase bass, sibilance and controlling this mainly with an acceleration limiter. I mean this is a common issue as who can sing without sibilance? Cliff Richard and Tim Finn are the only ones I know who can – amazing vocalists. Another issue is that there appears a push to make vinyl 'louder' as compared to vinyl from say, the 1970s, which would I think be correlated to mastering to CD when you could definitely go louder. This is probably OK on a 7-inch single, but on an album, this has its own set of challenges. If you make the first or outside track loud, which makes sense as it has the best radial curve, it is physically impossible to cut the inside track to the same level because physics won't allow it, so there could be up to a 2–3dB difference between these tracks.

Remastering Goodbye Yellow Brick Road *for CD 1995 (Tony Cousins)*

Sourcing the tapes (Tony Cousins)

The first thing we had to do with this project was source the correct master tapes. We did eventually get them because a guy called Bill Levinson from Atlantic was particularly keen on the project and was a friend of Gus, but initially there were some issues of tapes missing and/or the wrong ones received. This probably is a more common occurrence than you think. For example, I was working on a remaster for Roger Taylor (Queen drummer) on one of his earlier solo albums. I got the tapes from the record company and knew they weren't the right ones and so I asked them to go back and have another look. Now remember, this is Queen, a proper organisation with administration. Anyway, they finally located the correct tapes and they were found in the cutting room at Sterling Sound in New York where they were originally cut all those years ago. This was quite common; master tapes would sometimes be left in the cutting room where they were originally cut. I think the problem is even greater in a way now because people send you digital files and unless they are backed up, they can very easily go missing.

Another example is Elton John's *Tumbleweed Connection*. I was working on remastering stereo files for a guy called Greg Penny (who was working on creating surround sound mixes) and requested the master tapes. The record company was not being particularly helpful with my repeated requests and finally sent me a tape copy of the master and said this was all there was. Would you believe it, it sounded fantastic as it had hardly been played, and so we used that and I recall I didn't have to do much in terms of processing, mainly worked on getting the best digital transfer I could.

Once we had the correct master tapes for *Goodbye Yellow Brick Road*, we noticed there were a few problems with the condition – I like to refer to this type of condition as 'clapped out'. I know because looking at my notes after we did say 'Funeral for a Friend', I can see we added EQ at around 16–19kHz to try and give it a bit more air and life. This was further evidenced through the settings recorded on the floppy discs for the digital console we used.

Workflow and working with Gus (Tony Cousins)

The signal chain for this project consisted of firstly coming from the ATR 102 tape machine into the Dolby noise reduction system we had. When they recorded and mixed the album originally, they used the very first Dolby units made which were the A301 models. We didn't have an A301 model at Metropolis (we did at some time, but it was long gone) but we did have the 360 model which was introduced directly after the A301. We played the tapes through the 360, switching between regular and spectral noise reduction a few times, and realised that the sound was not quite right. We then managed to locate cards from the A301 model that we could insert into our 360 model, and this improved the sound tremendously. We therefore were trying to match the units as closely to the original recording as we could, and in other cases we try to source the original tape machines and heads as well to get the best possible transfer we can, as close as we can to how it was originally recorded and played back.

From the Dolby noise reduction unit, the signal then went into a custom mastering board built in house by John Goldstraw at Metropolis. This console housed Sontec equalisers and compressors as well as some Summit equalisers. There were also some equalisers built by Goldstraw, which were based on the Neumann equivalents in their SP 79 control consoles, plus a passive shelving unit based upon a Decca design. There were also available an Aphex Dominator compressor, a Maselec de-esser/limiter and an Ortofon SPL de-esser. Obviously not all of the components were used all of the time in remastering *Goodbye Yellow Brick Road*, but this was the selection on offer. From there the signal went through a PRISM 20 bit analogue to digital convertor and onto a digital limiter set at a small level between 1.5 and 2dB, never more than that. We were not trying to brickwall the signal, just raise the basic level in an effort to maintain a certain degree of dynamic range. After the limiter the sound was fed through to an eight into two Yamaha DMC1000 digital console, so the stereo signal would come up

FIGURE 3.2 Tony Cousins at Metropolis Studios [Photograph], received from Metropolis Studios

on each fader through to the stereo bus. Finally, the signal would leave the stereo bus on the digital console into a digital audio workstation (DAW) on the computer where the remaster would be recorded.

Gus would position himself at the digital console sitting on the main stereo pair. He would say things like I want half a dB there, or if it was a chorus add a half dB there, generally along those lines. He would make little tweaks on the fly, play them back, and if they were not quite right, we would do it again. I didn't know Gus that well before the project started, but if I said I thought some change was too much I would tell him, and he would either agree or disagree, but either way welcomed the input and feedback. It was a real joint effort. A lot of the time when remastering you are on your own in the room making your own decisions independently, so it was definitely different working with Gus.

One thing that became evident, because Gus knew the material so well as he produced the original recordings, was his intention to 'fix' things he may have been unhappy with in the original production. For example, he would say things like 'we're coming up to this part where there's no kick drum' or 'what's happened to the drum fill'. Gus could recall specific elements of the original recording where, for example, he felt a particular drum section or hit wasn't hard enough. This is backed up from my notes which suggest EQ in particular happening and being tweaked all of the time as Gus tried to 'fix' elements he felt needed fixing from the original release. Of course, the EQ manipulation was also a result of the rather poor physical condition of the master tape. Gus would make these fixes on the fly; we would listen back and then go again if he was unhappy. We did a lot of listening which really is the most important element of remastering.

We remastered for a CD release. Although the loudness wars had started (Oasis released the CD *(What's the Story) Morning Glory?* in 1995), we weren't really affected and tried to make this remaster as sonically acceptable as possible without the need for extra loudness. We edited it as a CD using software and 20-bit convertors and added index points. The gaps between the songs were copied from the original.

I think it is important to mention here that I know Ray Staff mastered the original recording. Ray is a highly regarded engineer, probably one of the most experienced in the UK and a lovely man, and much more technically astute than I am, so I knew from the outset that we would be remastering from a quality master.

Overall, I would say I was happy with the final outcome, although that is a difficult thing to answer. There is a part of me, from a pure mastering sense that would have been less happy with the amount of work we did, given I'm a great believer in 'less is more'. If Gus hadn't been in the room with me, I think I would have done less work as Gus did tend to try and 'fix' issues from the original recording, being so close to it. I still think it sounds amazing though, don't get me wrong, and I am not trying to lay any blame anywhere or on anyone. It just would have been a different experience and outcome working solo.

Specific tracks

'Candle in the Wind' (Tony Cousins)

We put a bit of a high top on it at 19kHz and then put this through a Sontech Compressor, which had a lovely sound. We saved these settings on the floppy disc and then came back later, set it up exactly the same except we boosted a bit more at around 5.5kHz. At roughly 2:30 into the song, when there is a bass entry, I added 1dB at 50 Hertz. We also added a subtle stereo spreader as Gus had gone home and I had a listen and thought it wasn't wide enough and just needed a little bit more life or presence in it.

'Benny and the Jets' (Tony Cousins)

For the end section on this track, on the left-hand side, we cut 3.5kHz. This is the part where there is a high harpsichord type sound in the left ear with a lower pitched version of the same sound in the right-hand ear and they perform a call and answer type movement. It would have implemented to subtly control and lessen the impact and energy around the higher midrange on this sound.

'Funeral for a Friend/Love Lies Bleeding' (Tony Cousins)

This track was done in sections because it starts quietly and builds up. The main reason it was done this way is that when you start very quietly and you're going into digits, you're not using all of the bit rate, so we started at a louder level. We also put a compressor across it and cut 6.8kHz on the synthesiser strings. As we did this in sections we didn't need to work on the fly and at the end I edited (or maybe someone else) it together as every section was different. The whole process got memorised, and then we went back, set it up exactly the same and then from section 2 there is slightly more additive EQ which means we need to pull the main fader down a little as there is more level going through.

From the acoustic piano entry, the introduction to 'Love Lies Bleeding', there is an EQ change at 0:58, then more EQ added at 1:34 and a cut at 3.4kHz at 2:39 – we must have put too much in the first time through. We also cut between 200–400k, a critical area for mastering, where it can sound 'boxy', to remove some of that unwanted tone.

Remastering *Goodbye Yellow Brick Road* for CD and vinyl in 2014

Goodbye Yellow Brick Road was remastered again in 2014 by Bob Ludwig at Gateway Mastering Studios, Maine, USA for a 40th anniversary special edition. Ludwig is a well-respected mastering and remastering engineer having won 12 Grammy Awards, and having worked with artists including Paul McCartney, Led Zeppelin, Queen, David Bowie, Madonna, Nirvana, U2, Jimi Hendrix and of course, Elton John (Gateway Mastering Studios, n.d.).

In an interview with Ben Robertson, Ludwig stated the following:

> When I got the call to remaster *Goodbye Yellow Brick Road* and then the subsequent remasters I was deeply honoured. These are indeed classics, and I gave them a lot of respect! In the past they were originally mastered for vinyl and cassette and, later, compact disc. For me, the goal of remastering these iconic albums in high-resolution digital for the future was to make them sound as good as I possibly could, yet paying close attention to the sound of vintage LP copies of the originals that we know were definitely approved by the artist before their initial release. That was the best pedigree for a reference.
>
> *(Rogerson, 2016)*

We can see from the quote above that Ludwig was honoured to remaster *Goodbye Yellow Brick Road* and approached the task in a respectful manner with regards to original production and the team responsible.

Thoughts on remastering

Ludwig approaches remastering in a different manner to mastering. For Ludwig, mastering is almost like starting with a blank canvas with a multitude of options available to the mastering engineer. Whereas remastering, according to Ludwig, has a previous version approved by both the artist and producer which must be considered. As he says:

> Remastering a catalogue album has different issues. One can buy a copy of the originally-issued LP and hear how it was first envisioned by the producer and the artist and know that sound was once approved. As our gear is definitely better than

> what was around for vintage LPs, one has to decide just how much 'better' is good, or perhaps if it should be left alone. There are a lot of variables. But at least one starts with a pre-approved musical direction, not the unlimited plethora of choices one has when approaching a new album for the first time.
>
> *(Rogerson, 2016)*

Ludwig was around during the well-documented 'loudness wars' in the 1990s where the CD format allowed for far greater levels of loudness than was physically possible on vinyl, and is thankful iconic albums from the past, such as *Goodbye Yellow Brick Road*, were spared from the 'hyper-compression' process. As he states:

> We're all afraid of the over levels, so people started inventing these digital-domain compressors where you could just start cranking the level up. Because it was in the digital domain, you could look ahead in the circuit and have a theoretical zero attack time or even have a negative attack time if you wanted to. It was able to do things that you couldn't do with any piece of analogue gear. It will give you that kind of an apparent level increase without audibly destroying the music, up to a point. And of course, once they achieved that, then people started pushing it as far as it would go.
>
> I always tell people, 'Thank God these things weren't invented when The Beatles were around, because for sure they would've put it on their music and would've destroyed its longevity.' I'm totally convinced that over-compression destroys the longevity of a piece. When someone's insisting on hot levels where it's not really appropriate, I find I can barely make it through the mastering session.
>
> *(Owsinski, n.d.)*

This is an interesting observation and is perhaps one of the reasons for the longevity of iconic albums from the past, why they remain so popular, and why we, as the consumer, want to own and listen to remastered versions that can fully optimise modern playback devices without destroying the dynamic range.

Comparative listening

I asked the case study participants to undertake comparative listening between a digitised copy of the 1973 vinyl first pressing, the 1995 remastered CD and the 2014 remastered CD for *Goodbye Yellow Brick Road* for a selection of songs and identify sonic differences they perceived.

Peter Kelsey

It was very interesting to listen to the three different mastered versions of the same material. I decided to listen to portions of songs to make it easier for me to compare. Obviously, the digitised vinyl version is at a lower level, but I also noticed that it had less high end and the crackling at the beginning of 'Funeral for a Friend' was very evident. I found this version the most 'warmest sounding' one, but this may be due to having less high end. I preferred this version overall but felt it could have done with a little more high end and level. This was noticeable on 'Benny and the Jets' but not as much on 'Candle in the Wind'.

For the CD remasters, I prefer the 1995 version although I found both remasters a bit harsh on aggressive parts of songs, especially 'Saturday Night's Alright for Fighting'. The 1995 remaster appears louder overall compared to the 2014 version. Furthermore, the 1995 CD version of 'Benny and the Jets' seems to be possibly distorted a little, although that may be due to adding too much high end.

David Hentschel

I have got one of the 'Goodbye Yellow Brick Road' CD remasters. I needed a digitised copy and this was an easy way to get it. I'm not a fan of digital, it still sounds brittle to me, particularly high-end sounds like cymbals, no matter how high the sample or bit rate is. There also appears to me too much presence in the vocals. It doesn't have the same smoothness that an analogue recording has.

Ray Staff

Only the people in the studio can know the sound of the original master. Listeners can only give an opinion of an altered master without reference to the original and within the limitations of their equipment and environment. How do you compare and judge the differences between mastering versions?

As a mastering engineer it is impossible to know exactly how the final audio will sound outside of the studio and in millions of different locations and differing equipment. This is a huge discussion that should include the many variables involved in comparing audio. This ranges from equipment tonality and specifications to listener bias and preferences. Equipment might measure well but often the designer or manufacturer will compromise on specification to achieve a sonic preference. Does that make it a valid piece of equipment for comparisons if it has a sonic bias?

It should also include sonic differences between formats for example vinyl and CD. When we mastered *Goodbye Yellow Brick Road*, it was only destined for vinyl. Later versions might have been required to encompass multiple formats with different technical specifications and limitations, and this might have required compromises being made.

When examining more recent versions it should also include the influence of modern digital releases that are often now part of the digital 'loudness' process. Some younger people have grown up with this sonic influence and might induce a bias or preference of how they enjoy listening to music. Will a master processed for digital loudness sound good on vinyl? Do you make multiple masters each optimised for the final format? There is no standard for the reproduction of music at home? So many questions.

Tony Cousins

Having listened to the three versions there are a few obvious things to say. The vinyl version lacks high frequency or is dull compared to both remastered CDs, as expected. And as I suspected, the influence of Gus Dudgeon being present in 1995 is of significance. It is almost uncanny how the CDs often sound very similar but rarely all the way through a track. I will list the essential differences, both CDs being 6db louder than the vinyl transfer with occasional internal differences.

'Funeral for a Friend/Love Lies Bleeding': The introduction on the vinyl and the 1995 is louder than the 2014 version until the band enter. Thereafter the 1995 version is slightly brighter than 2014 and the voice is louder.

'Candle in the Wind': The 1995 version is brighter than the 2014 one, with the voice louder and the track very slightly louder. The bass of the 2014 remaster is better defined, partly because there is less high frequency, it also makes it louder.

'Benny and the Jets': The digitised vinyl and the 2014 remaster are wider, especially the piano and audience. This is a detail that is difficult to understand. I have two hypotheses. An obvious one being that the tape was different, which I doubt. The other being that one of the things that Gus and I agreed on was that the width of the stereo image suggested reinforcing the centre image, however this should not pull in the stereo. The vocal on the 1995 remaster is louder.

Digital audio analysis

I applied the same digital audio analysis techniques as for The Beatles' chapter, having sourced and generated a digitised copy of a 1973 vinyl first pressing of *Goodbye Yellow Brick Road*, and both CD versions. I decided to focus on the three tracks which make up the complete first side of the double album, 'Funeral for a Friend/Love Lies Bleeding', 'Candle in the Wind' and 'Benny and the Jets', to determine how the mastering (and remastering) may vary from the outermost track of a vinyl album to the innermost track and to identify any consistencies (or inconsistencies) for a full album side in general.

Figure 3.3 is a screenshot that depicts the loudness/amplitude range over time of the respective waveforms for each release of *Goodbye Yellow Brick Road* within the Pro

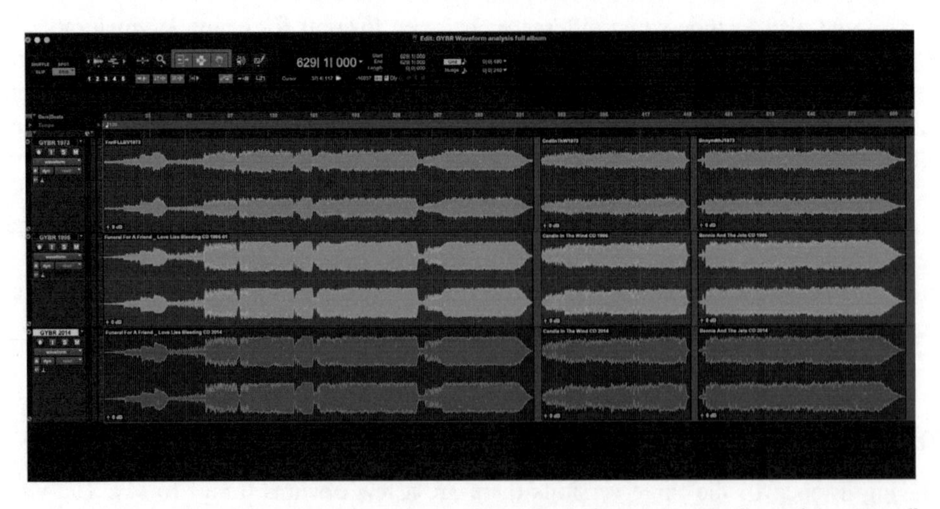

FIGURE 3.3 Waveform view of 1973, 1995 and 2014 versions of recordings from *Goodbye Yellow Brick Road*

Tools DAW software environment. The top horizontal track represents the 1973 vinyl transfer, the middle track depicts the 1995 CD release and the bottom track portrays the 2014 remastered CD release. As shown, each track consists of the individual songs 'Funeral for a Friend/Love Lies Bleeding', 'Candle in the Wind' and 'Benny and the Jets', in that order horizontally and aligned directly above and below each other. The visual representation of the 1995 and 2014 remastered CD releases clearly shows louder signals when compared to the 1973 digital transfer which is to be expected. Furthermore, the waveform shapes of both the 1995 and 2014 CD versions appear more 'block-like' than the 1973 version suggesting a more compressed signal overall with slightly less dynamic range. To explore these variances further, I measured left and right true peak level meter readings of the *Goodbye Yellow Brick Road* 1973 digital vinyl transfer, 1995 remastered CD and the 2014 remastered CD, similar to Barry's analysis undertaken on The Beatles (2013).

Table 3.1 represents the left and right true peak level values for the selected songs across the three versions. For the 1973 digitised vinyl version we can see that the true peak level measurements are quite strong for the first track on Side 1 'Funeral for a Friend/Love Lies Bleeding' and then decrease for both the second track 'Candle in the Wind' and final track 'Benny and the Jets'. This appears consistent with the physical limitations of vinyl when mastering, particularly that if you start off relatively loud on the first track, that it is physically impossible to maintain this loudness as the record progresses due to variances in radius. When we view the 1995 CD measurements, we can see roughly the same pattern is followed for Side 1, with the first track loudest, the second track softest and the third track in between. This would appear consistent with Cousins' description of Gus wanting to adhere closely to the original recording. By contrast, the 2014 version has the second track 'Candle in the Wind' with a higher true peak on the right-hand side than the last track which is slightly different. Additionally, both the 1995 and 2014 versions both show higher true peak levels across all tracks than the 1973 digitised vinyl version, which is to be expected not constrained by the vinyl format. However, both CD versions have measurements slightly above 0.00dB which would suggest digital distortion is evident.

Although this was helpful information in terms of identifying the loudest peaks across an individual track, it did not provide an average peak level for an arguably fairer loudness comparison. To explore these average variances in loudness, I also recorded the left and right root means squared (RMS) level meter readings to display

TABLE 3.1 True Peak level measurements *Goodbye Yellow Brick Road*

Song	*1973 digitised vinyl*		*1995 remastered CD*		*2014 remastered CD*	
	Left true peak	*Right true peak*	*Left true peak*	*Right true peak*	*Left true peak*	*Right true peak*
'Funeral for a Friend/Love Lies Bleeding'	-1.67	-1.78	+0.02	+0.16	+0.08	+0.14
'Candle in the Wind'	-3.73	-3.47	-0.97	-0.16	-0.20	+0.06
'Benny and the Jets'	-2.06	-3.28	-0.12	-0.09	+0.02	+0.01

the average level of loudness overall. RMS metering is useful for indicating if two or more songs are approximately the same loudness level (Owsinski, 2008).

Table 3.2 displays the RMS levels for both left and right channels across all three releases. The difference in average loudness between the 1973 digital vinyl transfer and both the CD remasters is quite significant, with the original master ranging between 4dB and 5dB, particularly on 'Candle in the Wind'. Although both CD remasters have greater average loudness levels than the original 1973 master, they are fairly similar with each other with only a difference of approximately 0.5dB across all four tracks. Furthermore, it appears that the 1995 CD remaster is softer in average volume of the left-hand side but louder on the right-hand side across all three tracks when compared to the 2014 CD remaster. Although these differences are less than 0.5dB, it is still a pattern that has evolved.

The next measurement undertaken was LUFS, a loudness measurement similar to RMS in terms of calculating average loudness, but which takes into consideration human perception of audio loudness and is an interleaved measurement (not separated by left and right).

Table 3.3 represents LUFS measurements across all three versions. It clearly depicts the 1973 digitised vinyl version as less loud than the CD versions by around 5–6 LUFS. It is also clear that in LUFS measurements, the overall audio perception of loudness, the 1995 CD remaster is slightly louder across all three tracks as compared to the 2014 version. This data appears to support Peter Kelsey's perception that the 1995 CD sounded louder than the 2014 release. Again, the same pattern for Side 1 is adhered to across all three versions, the first track 'Funeral for a Friend' is the loudest, the second

TABLE 3.2 RMS level measurements *Goodbye Yellow Brick Road*

Song	*1973 digitised vinyl*		*1995 remastered CD*		*2014 remastered CD*	
	Left RMS	*Right RMS*	*Left RMS*	*Right RMS*	*Left RMS*	*Right RMS*
'Funeral for a Friend/Love Lies Bleeding'	-15.93	-16.91	-11.20	-11.32	-10.95	-11.81
'Candle in the Wind'	-17.02	-17.86	-12.20	-12.14	-12.11	-12.86
'Benny and the Jets'	-15.87	-17.89	-11.47	-11.64	-11.15	-12.17

TABLE 3.3 LUFS level measurements *Goodbye Yellow Brick Road*

Song	*1973 digitised Vinyl*	*1995 remastered CD*	*2014 remastered CD*
	LUFS	*LUFS*	*LUFS*
'Funeral for a Friend/ Love Lies Bleeding'	-15.6	-10.16	-10.24
'Candle in the Wind'	-17.71	-12.21	-12.63
'Benny and the Jets'	-16.64	-11.03	-11.26

track 'Candle in the Wind' is the softest, and the final track 'Benny and the Jets' is in between. This appears to support the notion that both remasters respected and adhered to the original layout of Side 1 in terms of perceived loudness per track.

Table 3.4 depicts decibel measurements of dynamic range (DR) across the three releases of the three song recordings analysed. As displayed, the DR score for the digitised 1973 version, across all three songs, is the highest which implies that the DR on the original is greater than the DR of both CD remasters. This suggests that less compression may have been applied to the original as opposed to the remasters. Both CD masters portray the same DR scores for all tracks except 'Candle in the Wind', where the dynamic range of the 2014 release is slightly greater. Furthermore, it is evident that on the 1973 version, 'Candle in the Wind' has the lowest DR score of recordings from the original, whereas on the 2014 remaster, this song has the highest DR score compared against other tracks emanating for the 2014 remaster. This would suggest the 2014 remaster is not strictly adhering to the original release in terms of DR and the associated pattern across these three recordings on Side 1. The 1995 remaster appears fairly consistent in terms of DR scores across all tracks on this version, with only 'Benny and the Jets' slightly varying from the other tracks.

To further examine the characteristics of a reduction in DR and increased loudness inherent in the digital replicas, as opposed to the original analogue music artefact, the next measurement undertaken was frequency spectrum analysis. Similar to O'Malley's (2015) work, I was keen to examine the frequency spread of the three different *Goodbye Yellow Brick Road* versions in an attempt to identify where these dynamic differences existed and why they were made. It is important to note that the images shown in Figures 3.4–3.6 only provide a brief snapshot in time on the various frequency and volume levels across all three releases.

Figure 3.4 represents the frequency spectrum for the song 'Funeral for a Friend/Love Lies Bleeding', the first track on Side 1 of the album. The image was captured at around 1 minute and 16 seconds into the song where the synthesiser opens up to a 'polyphonic sound' across a number of notes at the same time (although each individual note was recorded separately). The light grey colour represents the 1973 digitised version, the 1995 remaster is medium grey and the 2015 remaster is dark grey.

It is evident that the 1973 vinyl transfer is less loud than the CD remasters, particularly the low-end ranges (70–150Hz) and at the mid to high end (4–20kHz). This appears to support both Kelsey and Cousins' sonic perception that the 1973 version

TABLE 3.4 Dynamic range measured in dB *Goodbye Yellow Brick Road*

Song	*1973 digitised vinyl*	*1995 remastered CD*	*2014 remastered CD*
	Dynamic range	*Dynamic range*	*Dynamic range*
'Funeral for a Friend/Love Lies Bleeding'	12	8	8
'Candle in the Wind'	11	8	10
'Benny and the Jets'	12	9	9

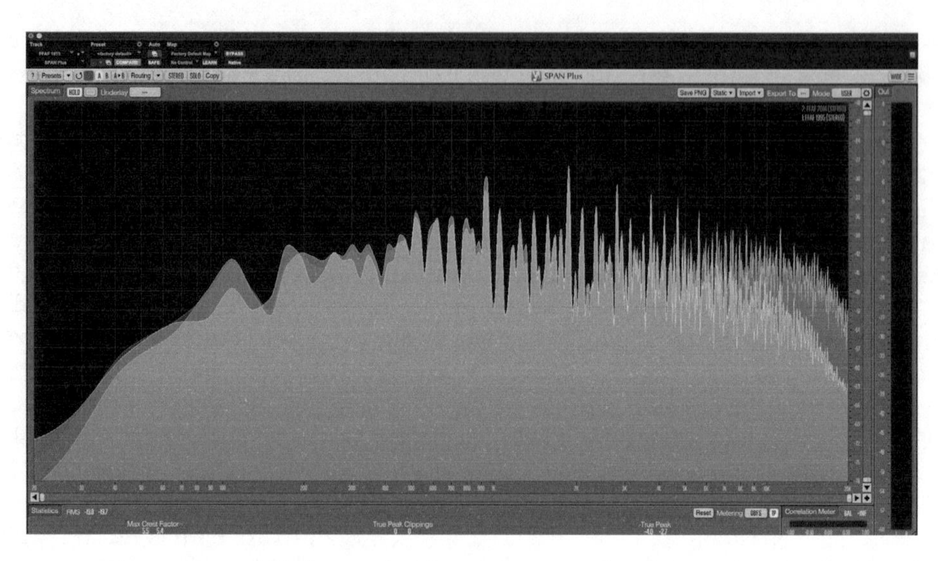

FIGURE 3.4 Spectrum analysis image for 'Funeral for a Friend/Love Lies Bleeding' – 1973 (light grey), 1995 (medium grey) and 2014 (dark grey)

appeared to have less high end. Furthermore, the lack of high frequencies in the 1973 version lends itself to support Kelsey's perception that the 1973 digitised vinyl version sounded 'warmer'. The 2014 remaster appears to have greater loudness from 35–80Hz, with the 1995 remaster displaying greater loudness predominantly in the 200–400Hz range. At the high end of the frequency spectrum, we can see greater loudness has been added for both remasters from around 5kHz until 20kHz as the shape starts to vary slightly to the 1973 version. The specks of dark grey indicate points where the 2014 remaster is slightly louder across those frequencies; however, the shape suggests frequency loudness is fairly even between both remasters in this region.

Figure 3.5 represents the frequency spectrum for the song 'Candle in the Wind', the second track on Side 1 of the album. The image was captured at around 38 seconds where Elton John sings 'seems' at the beginning of the first chorus. The light grey colour represents the 1973 digitised version, the 1995 remaster is medium grey and the 2015 remaster is dark grey.

It is evident that the 1973 vinyl transfer is less loud than the CD remasters across almost the entire spectrum, most noticeably from 4kHz and above where the gap is quite significant. However, the 1973 version does have greater loudness 20–35Hz than the 1995 CD remaster. We can see that the 1995 remaster is slightly louder than the 2014 version, particularly around the higher frequencies and around 200Hz. However, the 2014 remaster has greater loudness in the 20–70Hz range. This tends to support Cousins' perception that the 1995 remaster is brighter and louder than the 2014 one, however the 2014 remaster appears to have greater bass definition.

David Hentschel mentioned that overall he is not a fan of digital recording and/or playback as there appears to be too much presence in the vocals. This frequency spectrum appears to support Hentschel's sonic observation as we can see in the presence range (roughly 4.5–6kHz) there is a significant increase in both remasters.

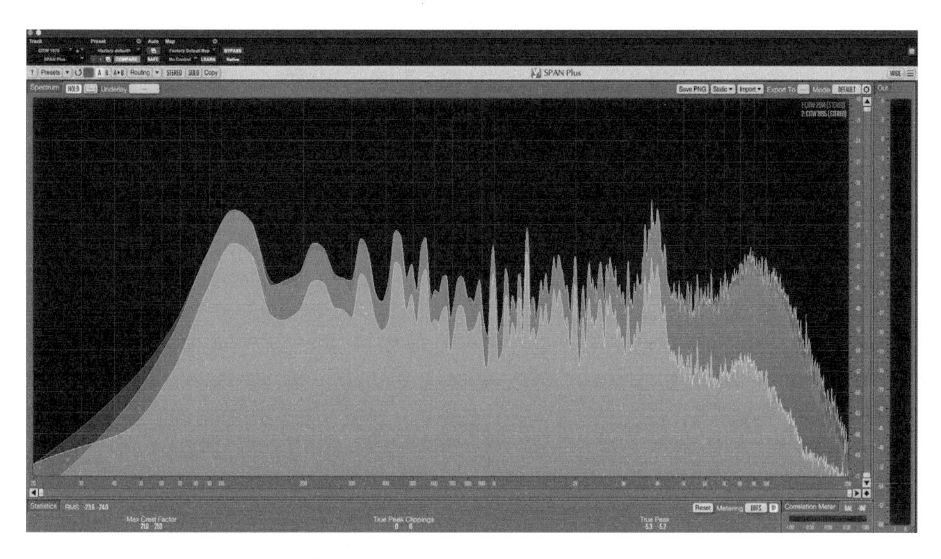

FIGURE 3.5 Spectrum analysis image for 'Candle in the Wind' – 1973 (light grey), 1995 (medium grey) and 2014 (dark grey)

Furthermore, both CD remasters boost in this range compared to the 1973 version where there appears to be more of a cut by comparison.

Figure 3.6 represents the frequency spectrum for the song 'Benny and the Jets', the third and final track on Side 1 of the album. The image was captured at around 28 seconds where Elton John sings 'hey' at the beginning of the first verse. The light grey colour represents the 1973 digitised version, the 1995 remaster is medium grey and the 2015 remaster is dark grey.

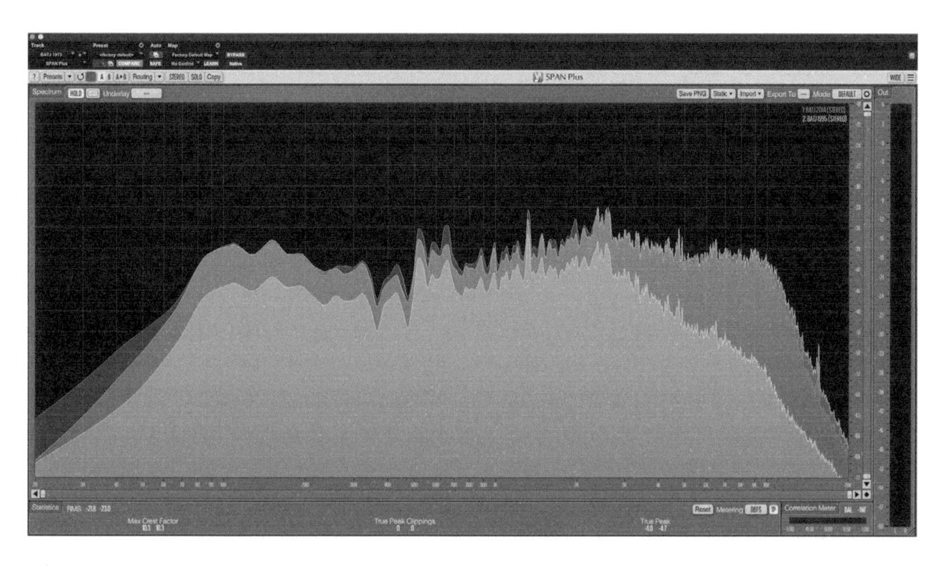

FIGURE 3.6 Spectrum analysis image for 'Benny and the Jets' – 1973 (light grey), 1995 (medium grey) and 2014 (dark grey)

It is evident that the 1973 vinyl transfer is less loud than the CD remasters across the entire spectrum and that it does not register any loudness above 18kHz. Similar to 'Candle in the Wind', it is evident that the 2014 CD remaster has the greatest loudness at the low end from 20–120Hz, suggesting more definition in this range. It also has the greatest loudness from 250–2.5kHz. The 1995 remaster has greatest loudness 150–250Hz and from 4.5–20kHz. Again, we can see a greater boost in the CD remasters from 4–10kHz where their respective shapes alter quite significantly from the 1973 version. This gap and change in shape lend support to Kelsey's view that the 1973 version could have done with a little more high end compared to the remasters. Cousins felt that the vocal was louder in the 1995 version and that the 1973 version was lacking high end and again the frequency spectrum image tends to support this view.

Cultural heritage

David Hentschel

There was a real value placed on vinyl back then and you had the whole physical package: the record itself, artwork, liner notes etc. People used to just collect, and they still do collect the artwork. Also, vinyl is analogue and when you listen to sound it isn't digital, we hear through our ears don't we. The whole thing about vinyl and playing it through a record player is that it is a mechanical process. Thank God vinyl has made somewhat of a reappearance. The quality of the pressings in the early days was actually very high. It was only once they started recycling vinyl that the quality of the pressings went down quite dramatically. I think they're better now though. If we put up with the scratches and dust of on vinyl, *Goodbye Yellow Brick Road* for example, to me it sounds way better than any digital version as that is how we designed it to sound.

To me there are two positive things about digital music production: editing and durability. Digital audio software is non-destructive so you can cut, copy and paste as well as undo which makes it a useful tool for editing. It is also durable in the sense that it doesn't degrade over time like analogue tapes and vinyl, so it makes sense to transfer them into the digital realm. However, I am quite anti digital remastering for the sake of remastering.

We designed and created the original recording of *Goodbye Yellow Brick Road* in a moment of creative time. You're in a little bubble, it's really buzzy and you have this image in your head of what you're going for. As Woody Allen said, the time when you're happy with your artistic creation is the time to give up. So, you never quite achieve it, and at the end of everything you always think there are always things you could have done better, but you have created this moment of creative time of everyone coming together and making this work of art. It is the coming together of talent at a moment in time and you master it put the final polish on it to make sure it sounds absolutely the best it possibly can. Then to have someone else come in later to remaster because the record company or whoever wants to make it louder, well, it completely blows out the window everything you tried to do when you were recording it, and that just makes me really cross.

> You can reprogram it through sampling, but don't call it the same thing you know. Put a fucking great label on the front of it saying this has been tampered with!

To me, remastering is destroying a moment in history, the original production. In a lot of cases, it's because the record company wants to hear it louder as it doesn't sound loud enough next to Beyoncé or have as much lower end. It's all simply synthesised bottom end to make a four-inch speaker sound like a 15-inch speaker – ridiculous!

Peter Kelsey

I do have a copy of *Goodbye Yellow Brick Road* on CD somewhere, so I guess it is a remaster. It seems fine to me to remaster as they've got more modern technology and can create something that is closer to the original sound. It doesn't bother me to have something remastered because I think what they're trying to do with that is to get back as close as possible to the original sound. However, I think there needs to be someone who can kind of sign off on it as far as adhering to the original intent of the recording. I mean, when you're mastering for vinyl, you're always making compromises for the vinyl and you do certain things in terms of EQ and so on so that will make it sound better on vinyl. Whereas, when you're mastering for a CD, you don't have to do that, you don't have the same limitations so you can make different choices. I mean really, a mastering engineer is there to just enhance a sound. If they're trying to do exactly what the original mastering was doing, which is to create that spectrum of all frequencies to make sure that all instruments are heard, to basically honour the mix, then that's what I'm looking for in any kind of remastering. Any kind of other manipulation, to make it sound like this because that is what we do now, no, I'm not interested in that.

Ray Staff

I have used modern digital and analogue equipment to remaster many projects. The decision to use any piece of equipment should be that you believe it achieves the best possible result at the time for the master you are creating. Why would you use an alternative just because it's analogue when it doesn't achieve the quality or result that's required? There are many things that are achievable in digital that are simply not possible in analogue. It is often desirable to use both analogue and digital to achieve the best possible result. The goal is the best possible result with the tools available to you.

Using digital is very desirable in many instances as it's the only possible method to achieve the result required. A bias against one type or method of processing would stop us achieving the best possible result. We have to trust the producer and engineers' judgement that by using any piece of equipment it is beneficial to and enhances the end result. Digital processing has improved considerably over recent years and along with higher sample rates digital is far superior today than with the early advent of digital formats.

I think there will always be someone who thinks they can improve upon an original. Does that individual have the same appreciation and understanding of the recording as the people originally involved in the project or are they superimposing their modern sonic and musical experiences onto an older recording inappropriately? Only their contemporaries will be able to judge the value of their work. Many of the original personnel are no longer around to make that judgement. For that reason, I think reissues should not totally surpass the original versions. Future generations may well want to investigate the differences and the value of newer versions.

Tony Cousins

To me the idea or process of remastering is to refer to the original as much as you can. The original vinyl not only has its physical limitations regarding loudness and bass, but it also represents the time and era in which it was made. When you are remastering you obviously refer to the original source material, however this tends not to reflect the first pressing, in the case of *Goodbye Yellow Brick Road*, because 50 years have passed and the first pressing is 'knackered'. To me the question is what element(s) have suffered because of this degradation over 50 years and does this need to be addressed? Furthermore, and this may sound a little off topic, but are there people around who have a sonic blueprint in their brain of listening to the first pressing in 1973 that they can recall? So, the point I am trying to make is remastering subjectively is difficult. Does the end result of the remastering process make *Goodbye Yellow Brick Road* sound better or worse? My view is that if I can make a recording sound more intelligible and show sonically more about what is there, then so much the better, and I will not fret over the fact that it does or doesn't sound like the original vinyl.

Bibliography

Buskin, R. (2011, October). Elton John: *Goodbye Yellow Brick Road. Sound on Sound*, accessed 10 June 2021, https://www.soundonsound.com/people/elton-john-goodbye-yellow-brick-road.

Discogs (2021). *Elton John – Goodbye Yellow Brick Road*, accessed 13 August 2021, https://www.discogs.com/Elton-John-Goodbye-Yellow-Brick-Road/master/30577.

Elton John (2014). Elton John – Goodbye Yellow Brick Road remastered & revisited (extended interview). YouTube, accessed 10 June 2021, https://www.youtube.com/watch?v=t4ODYfKrwqc.

Eltonjohn.com (2013). *Goodbye Yellow Brick Road* – recording, accessed 10 June 2021, https://www.eltonjohn.com/stories/goodbyeyellowbrickroad-recording.

Gateway Mastering Studios (n.d.). Bob Ludwig, accessed 7 March 2022, http://www.gatewaymastering.com/bob-ludwig/.

Greene, A. 2014. Elton John and Bernie Taupin look back at *Goodbye Yellow Brick Road. Rolling Stone*, accessed 10 June 2021, https://www.rollingstone.com/music/music-news/elton-john-and-bernie-taupin-look-back-at-goodbye-yellow-brick-road-205112/.

Johnstone, D. (2013). Davey Johnstone talks about Goodbye Yellow Brick Road, *accessed*10 June 2021, https://www.eltonjohn.com/stories/daveyjohnstonetalksaboutgoodbyeyellowbrickroad.

O'Malley, M. (2015). The definitive edition (digitally remastered). *Journal on the Art of Record Production* (10).

Owsinski, B. (n.d.). Bob Ludwig talks mastering and the loudness wars , *Bobby Owsinski Music Production Blog*, accessed 11 March 2022, https://bobbyowsinskiblog.com/bob-ludwig-mastering/.

Owsinski, B. (2008). *The Mastering Engineer's Handbook: The Audio Mastering Handbook*. Cengage Learning.

Rogerson, B. (2016). Bob Ludwig on remastering Elton John and why we still need pro mastering engineers. *MusicRadar*, accessed 7 March 2022,https://www.musicradar.com/news/tech/bob-ludwig-on-remastering-elton-john-and-why-we-still-need-pro-mastering-engineers-644489.

4

REMASTERING OASIS' *(WHAT'S THE STORY) MORNING GLORY?*

Introduction

Following on from the success of their debut release *Definitely Maybe*, English band Oasis released their next and most successful album of all time *(What's the Story) Morning Glory?* through Creation Records on vinyl, CD and cassette tape on 2 October 1995. Recorded at Rockfield Studios in Monmouthshire, Wales, and mixed (majority) and mastered at Orinco Studios in London, the album sold 347,000 copies in its first week and has since sold in excess of 22 million copies worldwide. The album was co-produced by the band's chief songwriter Noel Gallagher and Owen Morris, with Nick Brine working as recording engineer. According to Morris, 'they [the band] were very positive to be around. It was a lovely time, and for all its imperfections, the music was really, really good' (Buskin, 2012).

Leading into recording *(What's the Story) Morning Glory?*, Noel Gallagher describes not being sure how the album would turn out or how it would sound.

> I had no big sonic fucking I'm going to shift it this way or we're going to do this. I had some songs, enough for a record. We were a big successful indie band that were not going to lose the momentum of what we created. That was it. Nobody had any fucking idea what was going to happen. I wrote the songs, and I was more surprised than anyone
>
> *(Oasis, 2020a)*

> If you listen to the record, it's split into two halves. Half of the songs have got a second verse which I had written before I got here, and the rest of the songs have just the first verse twice, maybe a third time. And that was me getting in here going 'You know what, fuck it'.
>
> *(Oasis, 2020a)*

DOI: 10.4324/9781003177760-5

Recording the album *(What's the Story) Morning Glory?*

Rockfield Studios was founded in 1961 and is perhaps best known as being the place where Queen recorded the biggest selling pop single of all time in the UK (if you remove charity singles 'Candle in the Wind 97' and 'Do They Know It's Christmas?'), 'Bohemian Rhapsody'. It was also the world's first residential studio where artists and producers could stay in accommodation on site, and this appealed to Oasis frontman Liam Gallagher.

> There's a happy vibe on that album. I fucking love being in residential studios. You are a proper band aren't you when you are in there. Music is done different now. I do a lot of my stuff through email. You know, they fucking email you the drums and that these days. In 10 years' time you've got half a tune done. Where's the vibe in that?
>
> *(Oasis, 2020a)*

The camaraderie between the band members appears strong on this album given the short time it took to record and was most likely enhanced through the living on site arrangement. However, Noel Gallagher recalls that the recording sessions, and everyone living in close quarters, was not without its challenges.

> We'd (Owen and Nick) be at the desk, and everyone (the rest of the band) else would be sat on the couch, trying to do something and every now and then you'd turn around and say, 'Would you fucking shut up or fucking go back to the house?' We're trying to work and everybody's arguing about football.
>
> *(Oasis, 2020a)*

In an interview with *Sound on Sound* magazine, Owen Morris described the Rockfield Studios layout and his motivation to record there.

> Coming from Wales and having seen its name on plenty of records and it was a place where I'd always wanted to work. Rockfield has two studios: the Quadrangle and the smaller Coach House. When Oasis recorded there, the Coach House had a Neve VR console with flying faders, two Studer A820 multi-track machines, JBL monitors and a standard selection of outboard gear. A live room was situated to the left of the control room, directly in front of the control room was a drum area, and on the other side of the drum area was the main studio, with a couple of booths at the far end: a vocal booth on the right and a guitar booth on the left.
>
> *(Buskin, 2012)*

The band members and the main instrumentation/equipment used during the recording of the album is listed below as per the *Oasis Recording Information* website. There were also various other instruments used including a grand piano, tambourine, mellotron and Kurzweil electronic digital keyboard.

Liam Gallaher (lead vocals)

Noel Gallagher (guitars, vocals)

Guitars: Epiphone Riviera semi-acoustic; Gibson Firebird; Gibson Les Paul; vintage Fender Stratocaster, Epiphone Frontier acoustic and Takamine acoustic
Amplifiers: Marshall JCM 900 stack; Marshall Bluesbreaker; Orange amp; WEM combo and an early 1960s Vox AC30

Paul 'Bonehead' Arthurs (guitars)
Guitars: Epiphone Casino
Amplifier: Marshall JCM900 stack

Paul 'Guigsy' McGuigan (bass)
Bass: Fender Telecaster (Precision 61 reissue that has a Telecaster body shape)
Amplifier: Hi-Watt 200

Alan White (drums)
Gretsch kit

(Equipment used in the Morning Glory sessions, 2021)

Nick Brine was just 18 years of age when he took on the role of recording engineer for *(What's the Story) Morning Glory?* and a year later would work again with Owen Morris and the band on their follow-up album *Be Here Now*.

Nick Brine, recording engineer

Background (Nick Brine)

Music and football were my two major passions growing up in Monmouth, my dad being a former professional footballer, but at around 16 years of age I started leaning more towards music. I was in a band and studying for my GCSEs at school when a friend of mine, who also played in a band, told me he was leaving his job at Rockfield Studios, and they would be looking for someone to replace him. My dad ran a taxi firm in Monmouth and knew the studios well, having driven Robert Plant and others back and forth from there. I had a four-track recorder at home and was recording my band's rehearsals fairly regularly, so I had some experience. I called them up and had an interview. I finished my exams on the Friday and started with Rockfield on the Monday as a tea boy/tape operator.

My first session was assisting Andy Wallace (Nirvana, Rage Against The Machine and Jeff Buckley) on Sepultra. I spent about a month with him, and he really took the time with me to explain the industry, how I should behave in a studio and what I needed to learn first. From there, inhouse engineer Simon Dawson (The Stone Roses) took me under his wing and I worked my way up to engineering. The level of producers and engineers coming through Rockfield was phenomenal, so I was just like a sponge, watching these guys and picking up loads of tips and techniques. Rockfield also encouraged me to use the studios during downtime, so I would bring my band in to record. It was a great way to learn as I had to fault-find and figure out for myself how this was working and what to do. I started reading lots of manuals and books, but the best way to learn was getting your hands dirty. It was also on the cusp of the digital crossover, so it was an exciting place to be.

Working with Oasis (Nick Brine)

It was mental and lifechanging! I first met them when I was working on The Stone Roses album *Second Coming*, and I think they may have been recording *Definitely Maybe* around the same time. They came to the studio to hang out with Mani from The Stone Roses, so I got to know them then. They used to hang around the local pub waiting for Mani to finish recording so he could buy them a pint, as they were on small per diems at the time. When they turned up to record *(What's the Story) Morning Glory?* I already knew them quite well so that made it quite comfortable, considering I was 18 at the time.

It was a relatively fast project with 15 songs recorded in 15 days. There was a great sense of camaraderie within the band (despite what you may hear in the press) with all the band members present and supporting each other during sessions, working on and recording one song per day. Of course, there were pub visits and a few late nights listening back with a couple of parties thrown in, but overall, it was a very smooth process with a tremendous work ethic and commitment shown by the band towards this album. Noel had these songs mapped out and would play them, and we would knock them off one by one; it was quite methodical, and everyone involved was focused.

For their follow-up album *Be Here Now*, it was a very different experience. The process took seven months and involved moving between studios, press and fan intrusion and massive parties. It was more disjointed with a lot of stopping and starting, a few days working and then a few days off. They were also in the studio at different times, so the camaraderie that was there previously had tapered off. Noel would often do a lot of stuff on his own without the others there, and Liam would be on his own doing his parts. We still had fun, and it was life changing at 19 years of age to be involved in these sessions, meeting all these famous people hanging around the studio and doing all the crazy things we did. It was hard to take in, the level of intrusion and interruption and the type of press that was going on. It was insane. Obviously, I was part of the bubble, but a very small part. To be on the receiving end of what Liam was experiencing at the time would have been tough. I have loads of admiration for the way he handled himself.

Unlike *Be Here Now*, where Owen and Noel demoed the whole album to quite a substantial standard before recording it, the workflow for *(What's the Story) Morning Glory?* involved a lot less pre-production. Even though they had six weeks booked, they ended up averaging a song a day and it was all recorded in 15 days. So, there was no pressure to get a song a day recorded, that was just how it worked out. As it was recorded so quickly, there wasn't time to overindulge and do too many layers and overthink things which can happen sometimes when recording. For example, Liam's vocals sound raw and fresh as we did not spend two days recording them and/or editing them. We recorded them quickly, and as we were recording in analogue, we couldn't edit them that much anyway compared to working in digital.

It was about capturing performances mainly, which I am a big believer in, and I use that philosophy in my productions I do today. I try to get the camaraderie happening within the band and capture this, and ultimately their performances. I like to get everyone having a good time, happy and comfortable, and this is made simpler at Rockfield which is such an amazing place. It is inspiring. It is such a great place to

FIGURE 4.1 Oasis recording at Monnow Valley Studio in Rockfield [Photograph], photo by Michael Spencer Jones

walk into and you feel inspired by the fantastic recordings, artists and great engineers who have done work there before and I try to instil that in any band I work with there. As a producer, you have to pick when it's ready to go to the pub and/or work until 4.00am, so that the band are playing at their absolute peak and you're capturing the performances. I am a big believer in this, and a lot of great records are great because they have been recorded this way and they ultimately stand the test of time. And for me, that is why *(What's the Story) Morning Glory?* is such a great and revered album. They were on a high and enjoying the success of *Definitely Maybe* and there was that confidence present within the band and the songs they were recording, that this record would take them to another level, which it did. It was an amazing few weeks in my life, that's for sure.

Studio workflow (Nick Brine)

The first song we recorded was 'Roll With It', in which we set the band up in the studio to play together and recorded them live. However, after that song it was decided that they should approach things a little differently, which became the workflow we implemented pretty much daily for all other tracks on the album. Firstly, I would get to the studio at around 10:00am and Noel would come in and record a guide acoustic guitar and vocal to a click track. Noel and Owen worked out what the tempo would be for the song and its arrangement, and we would literally layer the song up from there. Alan White was a great drummer, despite being new to the band, and would record his part in two to three takes. Noel might make a few suggestions on types of fills and where to place them, but on the whole Alan would nail them. From there we would record Guigsy on the bass guitar, then Bonehead's rhythm guitar part and then Noel's main guitar bits. We would usually get to Liam's lead vocals in the early evening and then Noel and Owen would experiment with guitar solos, percussion and backing vocals. That was pretty much it for most days of the recording sessions.

I think this workflow worked well for Oasis and that's why it sounds fresh and has stood the length of time. There are a lot of recordings that take place that have all of the drums for every song for an album recorded first, then they move onto the bass, and then all the guitar parts and then finish with the vocals. I'm not a fan of this approach as I don't think it is fair on the singer. Towards the end of the booked sessions, you suddenly have a situation where you must record 12 vocals in three days which I think is gruelling and there is so much pressure. And then you have one day left to record all the backing vocals and/or percussion overdubs for the whole album. I don't mind working on two to three songs at a time, maximum, as sometimes you must do it that way, but I do like working on one song at a time.

Studio set-up (Nick Brine)

We set up Alan White's Gretsch drum kit in the drum room and placed an AKG D112 and either an Electro-Voice RE20 or a Sennheiser MD 421 on the kick drum. On the snare are Shure SM57s and there is a Neumann KM 84 on the hi-hats. We used either Neumann KM 56s or AKG C28s as the overheads and 421s on the toms. We also set up a stereo pair of Neumann U87s in the room corner towards the back and left the drum room door open with another U87 placed in the corridor for extra drum ambience, which is a signature sound of Rockfield.

For the bass guitar we used a 421 and a Neumann U47 FET on the amplifier and set up a DI Box as well. Owen Morris used to like using 421s on the guitar amplifiers along with SM57s, and we would also place either a Neumann U87 or TLM 170 as well to get that combination of sound from condenser and dynamic microphones. On Noel's acoustic guitar parts, using either his Takamine or his Gibson, we would use a combination of a Neuman U67, KM 84 and an AKG C414 and switch between them. Liam recorded all his vocals for the entire album on a FET 47 in the drum room. Usually, vocals would be recorded in the vocal booth or the dead area by the piano, but for this album we recorded Liam in the drum room because we wanted to capture a bit of natural air around the vocal being an ambient room, and to improve communication between Owen and Liam as they could have a line of sight and see each other. We kept all the microphones set up and in their positions for all the instruments during the entire recording, so it was easy for the band to just jump on their instrument and record when they needed to straight away.

For the drums, Owen added a distressor and dbx 160 compression on the kick and bass during recording to make them a little punchier, but not a lot. There was also a Urei 1176 compressor added to the ambient microphones, but again, not a lot. There was also a dbx 160 compressor added to the bass part. The guitars also sometimes had a little bit of compression added and then also went through these boxes called 'gain brains'. You don't find them in many studios, but Rockfield has some. Basically, these 'gain brains' create a small amount of harmonic distortion, and you end up with a tiny increase in gain and brightness and they sound really nice. This was all recorded to two Studer 24-track tape machines set up as a slave and master (we would fill up one first and then move on to the other reel) so we could use up to 48 tracks for mixing later. So, it was a totally analogue workflow controlled through a Neve 60 channel VR console and monitored through JBL4350s (main) and Yamaha NS10s (nearfield).

Recording 'Wonderwall'

'Wonderwall' was the biggest selling single off the album, across the entire Oasis catalogue and is ranked in the top 40 best-selling singles of all time in the UK official music charts (Official Charts, 2020). As mentioned previously, apart from the full band live performance and recording of 'Roll With It', all of the other album tracks including 'Wonderwall', followed a similar pattern of recordings allowing the band to record roughly one song per day, as described by Noel Gallagher.

> I would sit in the studio with an acoustic guitar and a click track of what I thought would be the pace of the song. Then I would sit and play the entire song all the way through without singing, of a song I am yet to write. And I would put on a guide bass, and a tambourine, and then one by one call the rest of the band down and say 'Right this is how it goes in this bit, you've got to do this, copy that bit, you do that bit, you fuck off to the chippy'.
>
> *(Oasis, 2020a)*

According to producer Morris, on the first day of the album recording sessions Noel presented two offerings of what would become 'Wonderwall': the version that ended up on the album and an alternate arrangement that was far more complex which Morris believed to be 'unnecessary and overcomplicated' so that version was scrapped. Two days later, Noel asked Liam which song he would like to sing, 'Wonderwall' or 'Don't Look Back in Anger', as Noel intended to sing one of them himself. Liam decided on 'Wonderwall', and they recorded the whole song that day apart from a small piano overdub (Buskin, 2012). Morris describes the process:

> First we got the drums and bass down by about midday or one o'clock … Then Noel overdubbed three acoustic guitar parts, after which Liam quickly did four takes of the lead vocal for me to compile. I used the Eventide DSP4000 pitch quantizer on his voice. If he was out on any lines, I'd just pull him in slightly, and everyone seemed happy with that. Having used it on *Definitely Maybe*, I used it on *Morning Glory* as well. Liam sang really well whenever I recorded him, and his voice was very open. So, he was doing takes very confidently and very quickly. He was so easy to record, and Noel absolutely loved his performance on 'Wonderwall'.
>
> *(Buskin, 2012)*

Nick Brine on 'Wonderwall'

We recorded the guide acoustic guitar track outside in the grounds on a wall about 8 feet high because the song was called 'Wonderwall', and that wall became known as the Rockfield Wonderwall. We set up Noel on top of the wall with his acoustic guitar and a Neuman U67 and KM 84, as well as an AKG C414. This is what you hear at the start of the album, this guide track with bird sounds and a slightly different rhythm to the final version of 'Wonderwall'. I remember we were recording and then we felt a few spots of rain coming down and the studio owner Kingsley Ward came out and asked, 'What the hell are you doing?'. I said, 'We are recording the guitar for Wonderwall' and I was just praying that it wouldn't rain as we had around £30,000 worth of microphones set up.

The string sounds you hear on the track are from a Kurzweil digital keyboard and there is also a Mellotron used, with Bonehead playing the root notes of the chords with the cello sound. These types of overdubs and additions to the songs were usually brought about by Owen and/or Noel experimenting late at night in the studio. Noel liked the idea of having atmospherics and segue ways on the album between tracks in general, and these generally came from Owen harvesting sample libraries. This allowed the recording of the album to progress and maintain momentum. For example, to record 'real' strings you would most likely have to wait a month or so to book an orchestra, get an arranger in, and then a few days off and go to Abbey Road Studios to record. This all helped with the flow of the album as everything could be completed in the studio while working through one song at a time. Owen and Noel would listen to a recording, think about what needs to be added, instruments and/or atmospherics, add it and then move on to the next song. Brilliant.

Recording 'Don't Look Back in Anger'

Noel Gallagher took inspiration for writing the song through listening to a recorded interview with Beatle John Lennon. As he states:

> On the *Definitely Maybe* tour, a guy working for our record company in New York had a brother who had a cassette of John Lennon in conversation in the Dakota Building just before he died. And in that cassette John Lennon says the words 'And then they'll tell you that the brains you had have gone to your head.' And that line stuck with me. And I thought 'I will fucking die if I don't shoehorn that into a song somehow.' That always stuck with me and is why the piano chords of *Imagine* are at the beginning, as a nod for that.
>
> *(Oasis, 2020b)*

According to Owen Morris, the workflow for this song was very similar to 'Wonderwall', and the piano introduction was very much a 'cheeky' addition and a bit of fun.

> 'When Noel got Bonehead to play it like *Imagine*, it made us all laugh,' he says. 'No one really cared. It was just funny. And as for the whole thing about his work being derivative of The Beatles, Noel was quite happy to be compared to The Beatles.'
>
> *(Buskin, 2012)*

Noel Gallagher took the lead vocal responsibilities on this track which was a move Owen Morris was not 100% onboard with at the start.

'I was never that convinced Noel should be singing it,' Morris admits. 'I didn't think he sounded as good as Liam. Certainly not back then. Liam was such a great singer. However, Noel sang that song, I comped it, and I also tuned it to ensure we had the vocal you hear on the record. Certainly, it worked, and people seemed to like it, but I've still always thought that Liam should have sung it. Having sung three lead vocals in three days, he (Liam) was quite happy to have the day off.'

(Buskin, 2012)

Reflecting upon the track's popularity and commercial success, songwriter Noel Gallagher laments being able to complete a song as a younger version of himself with no fear and confidence, to just follow your gut instincts.

It's really frustrating being in a musical collective when you're the only one who knows how it goes. If I had known now then what I know now about what 'Wonderwall' would become, or 'Don't Look Back in Anger', I never would have finished those songs. I'd still be fiddling around with it.

(Oasis, 2020a)

Recording 'Champagne Supernova'

Noel Gallagher's deliberation on the differences between writing and recording songs as a younger man compared to his current offerings is further exemplified through his song 'Champagne Supernova'.

Nobody knew how 'Champagne Supernova' would turn out because nobody had actually heard it. I mean it's seven and a half minutes long and nobody had heard it except me, and I wasn't really sure, as I hadn't finished it. I had the verse and the chorus, and the little guitar break in my head. I'd never played it obviously. You're just working on instinct as you don't second guess anything when you are young.

(Oasis, 2020b)

According to Morris, the recording of the guitar parts took about two hours. This included using an EBow on some guitar parts, a small electronic bow for guitar, which produces sounds not possible through normal picking or strumming. Apart from capturing this 'violin effect', Gallagher also recorded the traditional picking parts.

Noel would just sit in the studio next to his amps and play. He wouldn't come into the control room to listen to what he'd done until everything was finished. So, his guitar overdubs were all done fairly quickly.

(Buskin, 2012)

Liam Gallagher's vocals were recorded at one of the final sessions of the album. The reason behind this was that the vocals Liam recorded during the 'Champagne Supernova' session sounded not quite right to Morris. As he describes:

> We did half a dozen takes, but what happened was that the high note of the song – at the end of the line 'The world's still spinning around, we don't know why' – was kind of burning his voice out. So, he was getting croakier on each take, and by the time he got to the end he was sounding very Rod Stewart-y. I did a comp of the vocal and, bizarrely enough, Noel and Liam both liked it. But I didn't. I got him to re-sing it and we did it piece by piece. We did the first verse half a dozen times, and we did the ending half a dozen times. Then, once he'd completed all of the soft bits, Liam did the first chorus half a dozen times, followed by the same number of takes for the second verse and the other choruses, until he tackled the high part last.
>
> *(Buskin, 2012)*

The lead guitar part was performed by Paul Weller, most notably from The Jam, The Style Council and his solo work, who was well respected within the band. Oasis had recently been on the festival tour across the UK promoting *Definitely Maybe* and caught up with Weller and others. As Noel Gallagher recalls:

> I remember I sat on the back of somebody's tour bus playing Paul Weller the rough mixes of the album. And he went 'Fucking hell. I'll fucking play on that.' So, he came down and he thought he was going to play on 'Morning Glory', but 'Champagne Supernova' was crying out for a guitar solo because mine was a shocker. He played on that, and thank God he did.
>
> *(Oasis, 2020b)*

Working with Paul Weller (Nick Brine)

He's a great guy and there is so much respect for him from all of the Oasis guys. Paul definitely added a bit of buzz to the sessions, excitement, which kept the whole project rolling along nicely and at a high level. He came in to do his bit and once we set him up, we left him and Noel to it, and it worked out really well. I've also worked on another album with Paul at Rockfield and he is the ultimate professional and very talented.

Mixing *(What's the Story) Morning Glory?* album

You may assume that such a commercially successful record would have had a clear plan of how to achieve the final makeup of the recording. This would generally include, demoing all of the songs to work out compositional structures and arrangements, instrument tones and timbres, finalise lyrics and have reference mixes in place to help guide and produce the final sound. As Noel Gallagher states, this was not the case.

> Everybody goes on to me about the sound, the sound of that record. I can assure you that going in there, nobody had an idea of what they were doing. Because nobody had heard the songs because we didn't do any demos. We did 'Some Might Say' at Loco. I think we might have jammed 'Roll With It' in a sound check somewhere. The rest of it no one had heard before.
>
> *(Oasis, 2020a)*

The album was mixed at Orinco Studios (now Miloko Studios) in London by Owen Morris. However, before then a quick monitor mix and playback was done at Rockfield Studios so that the band and others could listen to the album in its entirety for the first time.

Nick Brine on mixing *(What's the Story) Morning Glory?*

Rockfield Studios (Nick Brine)

At the end of the recording process, we had a big playback in the studios and the band were there along with Brian Cannon (record sleeve artwork), Richard Ashcroft (frontman for The Verve) and the studio owner Kingsley Ward. Owen had completed monitor mixes and I lined them up, in the order that Owen and Noel thought it would end up, on a Digital Audio Tape (DAT) and we had a big playback. We had been listening to the songs back to back as they progressed, but to hear the whole album in its entirety, it just blew us all away. It was real hairs on the back of your neck kind of stuff and we knew we had a achieved something massive.

I remember Kingsley turning to them and saying in his Welsh accent 'You've bloody cracked it, boys. You've got five or six bloody big singles there and you're going to sell 10 bloody million copies!' Of course, it ended up being a lot more than that. I have spoken to Kingsley since and he told me that the only other time that he was so confident of a record being huge, and remember he was involved in a lot of big records from the 1960s, was when he heard 'Bohemian Rhapsody' by Queen in the studio for the first time. And I think everyone in the room knew the way the country was and the politics that this was going to lift them to another level. It was just the perfect record to make as it was different to what everyone was expecting. It was melodic, the acoustics sounded great and had string parts. It was brilliant.

Orinco Studios (Nick Brine)

I wasn't directly involved in the mixing process at Orinco Studios, but I was present and did observe the mix sessions. I'd started working with Owen from that point anyway and was there just to help where necessary. Orinco Studios had their own studio assistant who was well versed in how the studio worked, and they would be responsible for plugging in and setting up all the outboard equipment and getting the room ready for mixing sessions. My job was really to look after the band which included driving them to and from the hotel and maybe taking Liam to the pub.

Like recording, this was also a quick and methodical process. Owen would create a mix and then Noel would arrive, have a listen and then maybe tweak a few things. The band would then come in for a listen, maybe add a few suggestions and then these would be made in the morning when Owen had fresh ears. I recall there were not as many late nights as when we were at Rockfield, maybe a couple of nights in the middle, as the focus was to keep Owen's ears as fresh as possible.

I don't remember Owen using reference mixes at Orinco or Rockfield. There are certain producers who listen to a lot of reference material in the room before they start work, but Owen was never really like that. He obviously was listening to a lot of records where he possibly wanted to go with the mix as opposed to getting used to the room. But the mix didn't really change huge amounts from the monitor mixes that we left Rockfield with. It was interesting for me, as a young kid, to hear what had been recorded at Rockfield, and how this would sound mixed in a different environment. It was a great experience to hear what was different about it, and maybe something I thought hadn't worked when we were at Rockfield and that it worked perfectly when we were listening to it in a mix environment. I think one of the main goals was to not lose that fantastic feeling and excitement we felt when listening to it at Rockfield on the big JBLs as Owen likes to monitor extremely loudly, deafening, and I believe he achieved this.

Mixing 'Wonderwall'

Owen Morris did two mixes of 'Wonderwall' – the monitor mix at Rockfield Studios and the 'proper' mix at Orinco Studios – and chose the Rockfield version to go on the final album. As he states:

> I chose to use the Rockfield mix. It sounded better to me. I'd put extra strings into Rockfield's echo chambers and you can hear the swirling by the end. Noel approved what I was doing, but he also let me choose and change mixes until I was happy. No one else was involved.
>
> *(Huggins, 2011)*

Mastering *(What's the Story) Morning Glory?*

The album was mastered at Orinco Studios by Owen Morris and then the mastered tapes were sent to Nick Webb at Abbey Road who transferred them to vinyl, CD and cassette tape. Owen describes the mastering process in an interview with David Huggins published on the Oasis Recording Info site:

> A year or so before I mixed *Definitely Maybe* for Oasis, I re-mastered an album I engineered for Johnny Marr and Bernard Sumner: the first Electronic album (you will find the re-mastered version I did on iTunes). I hired in an Apogee A/D which I'd read had a 'soft limit' feature. Given that I had no confidence in the sonic integrity of my mixes I had decided that I would attempt to use VOLUME (i.e., quantity rather than quality!) as my rather blunt tool. The soft limit feature on the Apogees allowed me an extra 6db or so gain before distortion.
>
> When I'd finished mixing Oasis's first album a year later, Oasis's management allowed me to master the album myself. Again, in Johnny Marr's studio, I mastered from half-inch tape masters, though a TC Electronics four band stereo parametric equaliser and then into the Apogee A/D with soft limit on and then into an Apple Mac running a basic stereo Sound Designer program for editing and assembling. Then I copied the edited stereo file to a DAT tape, which was then sent to a mastering house for ONE TO ONE digital cloning to have P Q coding etc. added, as needed for CD production. At NO STAGE did any other mastering

engineer add compression or do ANYTHING other than copy ONE to ONE MY digital master.

I would then receive a test CD from the mastering room so I could check that the 'mastering engineer' (person who copied a tape and added P and Q info based exactly on the timings I had specified) hadn't fucked up and attempted to change anything.

A year later, when I mastered *Morning Glory*, the only difference was that I used a pair of Neve 1081 EQ's instead of the TC's. This gives the album its distinctive sound.

Only later, on *Be Here Now*, did we take my mixes to be mastered by someone else, Mike Marsh at the Exchange.

I just got lucky with the Apogee A/Ds. *Morning Glory* was massively about the Neve 1081s which we'd hired in from some hire company in London … they had input gain pots before the normal Neve 1081 step gains (which are in 3 dB? steps, possibly more). This allowed me to get the EQ's right on the edge of distortion for each track. With both *Definitely Maybe* and *Morning Glory*, I used mastering as a tool to help my not very great sonically mixes to sound OK in the outside world.

(Huggins, 2011)

Nick Webb on mastering *(What's the Story) Morning Glory?* vinyl

Background (Nick Webb)

Nick had a long and distinguished career at EMI and Abbey Road Studios, working as a tape op and recording engineer on works by The Beatles (and some of their solo albums), Pink Floyd, Phil Spector and The Pretty Things. From there, Nick moved into mastering and remastering for artists including Paul McCartney, Queen, Deep Purple, The Beach Boys, Pet Shop Boys, Iron Maiden, Duran Duran and Tears for Fears well as orchestral works featuring Andre Previn, Sir Adrian Boult and Sir John Barbirolli.

Mastering Oasis (Nick Webb)

Barry Grint, a fellow mastering engineer from Abbey Road, had done either an acetate or vinyl test pressing of *Definitely Maybe*. At that stage it was to be a two-sided vinyl release. I received a call from Creation Records saying they were unhappy with the cut as it was too quiet, which I found odd as we always strived to cut records as loud as we can within the physical limitations of vinyl. I checked Barry's cutting notes and found out that each side of the album was 25 minutes, which is really long. I told Creation Records that the problem was the length and that a solution to achieve a louder cut might be to release it as a double album. They agreed. I ended up cutting the vinyl for *Definitely Maybe* and from there, they asked me to cut *(What's the Story) Morning Glory?*

Owen Morris (or somebody from the company) sent production master DATs over but it was unclear by the labelling whether I was to add EQ or not. Anyway, I played them, and they sounded loud and fine, and I thought 'I'm not touching these'. Oasis had already done quite well by this stage so I took the time needed to ensure I produced a cut as loud as I could without compromising elsewhere. This involved sending the recording from the DAT through the EMI TG console and then onto my Neumann lathe. Obviously coming from a DAT to a vinyl cut, you've got to turn it down. You can't cut it at say +6–7dB for an album as it is not physically possible, so you need to set your optimal level. We would keep notes on this as well as how many microns you were cutting out, where you had lateral deepening and how much space you had between the grooves. This was particularly useful if clients needed a recut at a later stage and we could simply reference the cutting notes. Anyway, I made a test pressing and sent it over and they were happy with it.

For the CD release, because it was so good in terms of already being at an optimal level (I don't know how many loudness boxes it had been put through) I just did a direct transfer. This involved transferring the DAT recording into the Sonic Solutions DAW on a Mac where I PQ'd it (adding codes that run alongside the audio data containing metadata for track start, pause, index, table of contents, ISRC and UPC codes used to create a production master or Redbook CD from which commercial copies would be produced from). Again, I sent over a copy to the record company and Owen, and they were pleased with the result.

They were an interesting band, Oasis. I remember they were in Abbey Road Studio 2 recording their follow-up album and you could often hear them as they were so loud. My colleague worked as a tape op on the sessions and told me that the band often didn't turn up when they were supposed to, but Noel was always there making sure everything was alright and organising recordings. To me, this was similar to Paul McCartney's position during the later stages of The Beatles. So, I have a lot of respect for Noel, in particular his work ethic.

Loudness wars

The brick wall mastering technique that Owen Morris developed through mastering the band Electronic's first album in Johnny Marr's studio which he then applied to Oasis releases *Definitely Maybe* and *(What's the Story) Morning Glory?* led to a battle of loudness with other releases of the time, most notably Britpop rival Blur, which was termed the 'loudness wars'. The CD format and the 'soft limit' function on the Apogee A/D convertors allowed Owen to produce CDs that were significantly louder through hyper compression, and other mastering engineers would soon follow. As he explains:

> I wasn't technically that adept, but the theory was loud and proud, and I did a similar thing on *Morning Glory*. Then, after that, I thought I'd better stop. There's only so loud you can go!
>
> *(Buskin, 2012)*

Nick Brine

The loudness wars were around as I remember certain CDs, when you played them or they came on the radio, they took people's heads off. It doesn't matter so much now obviously as everything is limited and you can only go so loud anyway before it sounds awful. Overall, in the 1990s, I think compression went too far and levels were recorded at levels that were too hot, but sometimes this worked, as is the case for *(What's the Story) Morning Glory?* I think that was Owen's intention from the beginning to record it hot, and mix and master it as loud as possible, and it works. There are other albums from that period which followed a similar intention that to me, did not work. I would question the loss of dynamic range.

I remember working with other artists and they would complain and say, 'I've played it against this, and it's not as loud' and you would tell them that when it gets mastered it will be louder and so there was this constant battle. The best engineer I ever worked with regarding recording at hot levels was Roy Thomas Baker who engineered the Queen records. I did a record with him with the band The Darkness, and because of his recording levels I thought I am going to have so much bleed, particularly from the drums. But he made it work and pointed out the amount of natural compression we were getting off the tape machine. He was the best.

Nick Webb

There has always been a loudness war, long before CDs came about. Achieving loudness on vinyl was a huge issue, particularly for singles. I remember a band came into the mastering suite, this was in the days of vinyl and tape cassette, and said to me 'We bought the Fleetwood Mac album, and we want to be as loud as that'. I thought OK, so we played the album in the studio, and it was louder, but I said 'Hang on, listen to it. They've rolled all of the bass off and sacrificed a lot of bottom end. I can do that if you like' and of course the band wanted to maintain the bass, so I cut their record at a level lower than Fleetwood Mac, but with more low end present.

I think when CDs arrived, loudness got a bit out of control. There was a heavy metal grunge band I was working with, and one of the guys came in with an image of the waveform of one of their recordings and it was just a brick wall. There was not an ounce of air there at all and it sounded awful, like a fly in a jar. When you compress so heavily, you lose some of the 'goodness' and definition, and some of the bass as well. I always explained this to clients and gave them a choice of how they may like me to proceed with the mastering. Some people listened and opted for a lower level with greater definition and dynamic range, and others preferred it to be really loud and squashed. I don't think a mastering engineer is there to say this is the way you need your album or your single to sound unless they ask.

Remastering *(What's the Story) Morning Glory?*

The album was remastered in 2014 as part of an expanded box by Ian Cooper and Owen Morris (Andy Hippy Baldwyn remastered some of the outtakes) at Metropolis Studios in London. It was released on vinyl, CD and tape cassette through Big Brother Records, set up by Oasis, and featured the original recording along with outtakes and alternate mixes.

Ian Cooper on remastering Oasis

Background (Ian Cooper)

I was brought up on a farm in Somerset and remember seeing a film crew and decided I wanted to be a cameraman. I started a film technician course which included electronics, vision and sound at Ravensbourne College in London. Unfortunately, I wasn't strong on the theory side (math and electronics) and started applying for jobs at recording studios. I had an interview with PYE at Marble Arch, and the manager asked me if I had any questions. I had just learnt a little bit about disc cutting so I asked them what cutting styli they used, rubies or sapphires, and he told me they used rubies. I then said, 'What do you do with them after you are finished with them?' He replied, 'We throw them away.' I then asked, 'Why don't you sell them?' At that point he stormed out of the office down the corridor, I heard him shouting at the Studio Manager, and he came back into the office and said 'Right, you start on Monday!' and this was in 1971.

I learnt to cut records at PYE, mainly Irish, Reggae and various other smatterings of music, using an old American Scully valve lathe. I also had a valve Ampex tape machine (that had no clock on it) which had a habit in the evenings of slowing down resulting in a change in pitch, so it was quite challenging. I was told that my job was to transfer the master tapes, and if anything was wrong it was the engineer's fault. I learnt through experimentation that you could vary the sound through applying EQs and compressors and shortly after I was upgraded with a new solid-state Neumann VMS 70 lathe with an SP272 console, built-in Neumann ash tray and SAL 74 cutting amps which made life a lot easier. However, I learnt that there was a difference in sound between transistor and valve systems, and that valve was quite good at disguising distortion.

It was great fun working at PYE and I got to do some interesting things. I cut some early experimental dbx noise reduction encoded records, when you played them back, the noise was gone. It was quite eerie. Furthermore, I also did some cutting for the Sansui encoded QS quadraphonic encoded format. I also started to do some work for a 'hippy' company from Oxfordshire. They weren't particularly well received by my colleagues, for being different I guess, but I quite liked them. That 'hippy' company would later become Virgin.

Richard Branson (Ian Cooper)

I moved from PYE to Utopia in 1976. They had a purpose-built room with upgraded equipment, shag pile carpet and huge fish tank, so it was wonderful. I had many clients including Virgin, who followed me from PYE. One day I met Richard Branson for lunch on his canal boat, and he was asking me questions about cutting equipment costs and setting up a mastering suite. Anyway, by the end of lunch I was working for Virgin at their new facility Town House Studios (known as Townhouse) alongside Tony Cousins, who I had known for a while. Tony and I eventually left and with Tim Young set up Metropolis Mastering (Adele, Robbie Williams, Amy Winehouse, Ed Sheeran, The Killers and Led Zeppelin) where I worked on remastering the Oasis back catalogue.

Mastering Oasis (Ian Cooper)

I mastered Oasis' compilation album *Stop the Clocks* in 2006. The record company sent over 23 boxes of tapes and Noel Gallagher came down to Metropolis quite regularly to identify which tapes were the masters. On the tape boxes there were comments such as 'this is the master' and 'no, this is really the master' which wasn't very helpful, but the project worked out well. The tapes sounded so good I don't remember really doing a lot to them. I recall the band coming down to listen to the whole completed *Stop the Clocks* in one sitting. I set them up in the room with tea and coffee and showed them the volume control and left them to it. I told them I generally listen at level 4.5 level, but, knowing their reputation for volume, told them it went up to 12. Anyway, the room was thundering along, and they staggered out and Liam said 'I don't know about you, but I think my ring piece has just fallen out of my arse' which I found quite funny. They had got up to 6.5 level. I also worked on a few singles and mastered Noel's first two albums.

I rang their office one day to let them know I was retiring, and their response was to get everything remastered before I left. I had a chat with Owen Morris, and we decided to do one album per day and that's how it happened. The workflow we followed started with identifying the correct master tape. We would load the tape onto an Ampex ATR tape machine (in some cases the master tape had to be baked first) that we had in the studio. The ATR had the advantage over the Studer tape machines of having dual playback heads (one transistor and one valve) and could also handle 14-inch spools. From the ATR, the analogue signal chain would continue with (and obviously we didn't necessarily use all components all the time) an EQ stage (Maselec and Sontec); then a compressor stage (Shadow Hills and Maselec) and then onto what Tony Cousins and I termed the 'muck spreader' which had bass filters, top end filters, mid/side function and a stereo width control. From there we used PRISM A/D convertors at 24bit 96kHz into Sadie, which we used as a tape machine. If the final product was CD, then we sampled down to 44.1kHz.

We didn't use any digital processing apart from conversion at the end. When I am working, I am always listening, and simply select the frequency and turn the knobs until I think it sounds good. Whereas with digital, you are looking and listening which I find distracting. I do recall we had to work from a DAT for one of the albums, but I can't remember which one, which didn't sound quite right. In the digital domain Owen used a TC Electronic Maximiser to limit the hell out of it, but after that little episode, we returned to analogue.

The tapes sounded great which was a tribute to the band and Owen. I think it was good for Owen to go back through history and realise what a great job he had done with the band first time round. Apart from minimal processing, there really wasn't a lot to do. Unfortunately, my notes have been filed away or thrown out so I can't give too much detail but generally we followed the same formula. We would almost always put a filter on the bass, maybe add or cut 1dB here or there across the mid and high frequencies and a little bit of 'muck spreading'. Owen was a pleasure to work with and the project was great because there wasn't a lot to do which is a credit to all involved in the original production.

Working with Noel Gallagher (Ian Cooper)

I would run off copies of each remaster we did and send them to the record company who would forward them on to the band to listen to. I don't recall Liam ever coming

FIGURE 4.2 Ian Cooper and Noel Gallagher [Photograph], received from Metropolis Studios

down to Metropolis during these sessions, but Noel did every so often. He was easy to work with. He would come in and we'd have a bit of a chat and then he would listen to the remaster while having a cup of tea and making notes. They may have included suggestions around fading or tone. He'd then often go for a walk down the Chiswick High Road while I made the changes he suggested and got it ready to play back to him. He'd come back, maybe have some lunch, and have a listen to the version and then take a copy away with him. That was it!

Comparative listening

I asked the case study participants to undertake comparative listening between the original 1995 CD release and the 2014 remastered CD of *(What's the Story) Morning Glory?* for a selection of songs and identify sonic differences they perceived.

Nick Brine

If I'm going to sit and listen to the whole album, I still enjoy listening to the original CD version as there is just something about it. The remastered version may be slightly wider with more space in the mid frequencies and a little shinier overall, but I don't think it is different enough for me to go 'wow', which makes me gravitate towards the original. Of course, listening on different platforms will highlight some differences as well. I prefer the original on CD and Spotify, but possibly the remaster for iTunes or YouTube.

'Hello': This song has quite a mish-mash of sounds inherent within it and I would say the remaster is more clear and less cloudy. But there is just something about the original, it just works for me, and everything is still sat where it is supposed to be.

'Wonderwall': I prefer the original version for this mainly because of the way the vocal sits – I just love it.

'Don't Look Back in Anger': I prefer the remaster for this as they removed some pointy aspects from the original, which I quite like.

'Some Might Say': I find that the remaster is tightened up and less floppy at the bottom end so I would choose that version.

'Champagne Supernova': I find the remaster is crispy and open which is nice; however, I like the honky sound of the guitars on the original where the mid frequencies are stronger.

These observations are made when you listen to the tracks individually, but when you listen to the whole album all that goes out the window because I prefer the sound of the original – it just flows.

Digital audio analysis

I was able to compile 16bit/44.1kHz WAV files from both the 1995 original CD and the 2014 CD remaster so that they are directly comparable within the same format. As the tracks emanated from CD as opposed to vinyl, physical limitations and differences associated with vinyl were not applicable. I chose the tracks 'Wonderwall', 'Don't Look Back in Anger' and 'Champagne Supernova' based on duration, timbral and dynamic differences between them.

Figure 4.3 is a screenshot that depicts the loudness/amplitude range over time of the respective waveforms for each release of *(What's the Story) Morning Glory?* within the

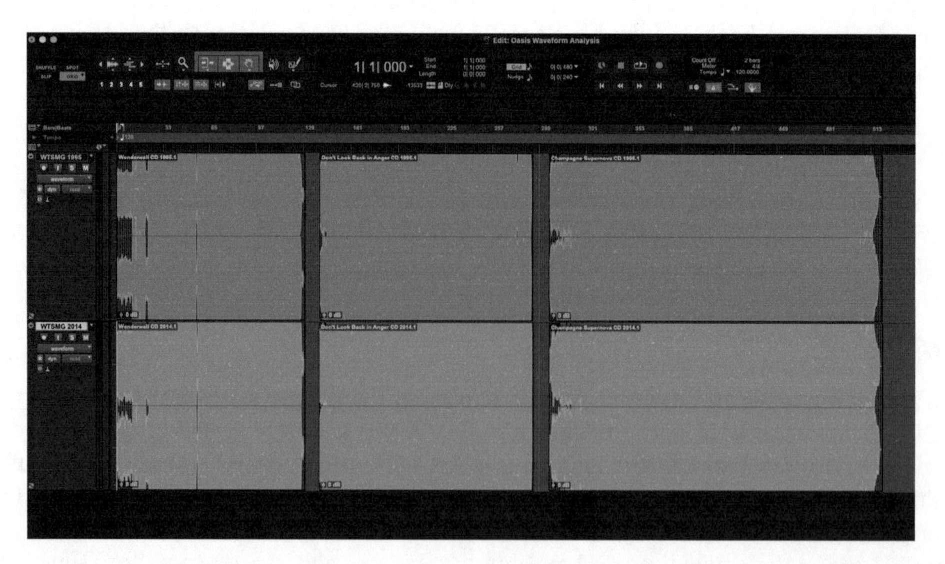

FIGURE 4.3 Waveform view of 1995 and 2014 versions of recordings from *(What's the Story) Morning Glory?*

Pro Tools digital audio workstation (DAW) software environment. The top horizontal track represents the original 1995 CD release, and the bottom track portrays the 2014 remastered CD release. As shown, both tracks consist of the individual songs 'Wonderwall', 'Don't Look Back in Anger' and 'Champagne Supernova' in that order horizontally and aligned directly above and below each other. The visual representation of both versions clearly depicts a 'brick wall' shape which suggests hyper compression is present, although this appears slightly less on the original version. To explore these variances further, I measured left and right true peak level meter readings of the *(What's the Story) Morning Glory?* 1995 mastered CD and the 2014 remastered CD, similar to Barry's analysis undertaken on The Beatles (2013).

Table 4.1 represents the left and right true peak level values for the three selected songs across the two versions. For the 1995 version it is evident that 'Wonderwall' and 'Don't Look Back in Anger' share similar peak levels over 0dB, with 'Champagne Supernova' recording the highest peak levels, suggesting digital distortion is present and consistent to levels achievable on the CD format, as opposed to vinyl. The 2014 remaster follows a slightly different pattern to the master, in that 'Wonderwall' has the highest peak levels and 'Champagne Supernova' the lowest, and the only track with a negative reading. 'Don't Look Back in Anger' registered similar levels across both versions which appears consistent with Cooper's claim that some tracks had minimal intervention. It is obvious from these readings are from CD releases, as they would be difficult to attain from vinyl.

Although this was helpful information in terms of identifying the loudest peaks across an individual track, it did not provide an average peak level for an arguably fairer loudness comparison. To explore these average variances in loudness, I also recorded the left and right root means squared (RMS) level meter readings to display the average level of loudness overall. RMS metering is useful for indicating if two or more songs are approximately the same loudness level (Owsinski, 2008).

Table 4.2 displays the RMS levels for both left and right channels across both releases. It is evident that the 2014 remaster is fairly consistent in average loudness levels across all tracks. On the 1995 original 'Wonderwall' is significantly quieter in average loudness compared to the other two tracks, and overall, this version has the greatest variation of average loudness, with a 3dB difference between 'Wonderwall' and 'Champagne Supernova'. Also, the original CD version of 'Wonderwall' is 2dB quieter in RMS compared to the 2014 remaster, and the original 'Champagne Supernova' is

TABLE 4.1 True peak level measurements *(What's the Story) Morning Glory?*

Song	*1995 mastered CD*		*2014 remastered CD*	
	Left true peak	*Right true peak*	*Left true peak*	*Right true peak*
'Wonderwall'	+0.04	+0.09	+0.29	+0.39
'Don't Look Back in Anger'	+0.03	+0.06	+0.05	+0.07
'Champagne Supernova'	+0.82	+0.36	-0.12	-0.12

TABLE 4.2 RMS level measurements *(What's the Story) Morning Glory?*

Song	*1995 mastered CD*		*2014 remastered CD*	
	Left RMS	*Right RMS*	*Left RMS*	*Right RMS*
'Wonderwall'	-8.80	-8.64	-6.06	-6.17
'Don't Look Back in Anger'	-6.64	7.12	-6.68	-6.53
'Champagne Supernova'	-5.85	-5.77	-6.68	-6.82

1dB louder than the remaster. This is interesting as it challenges the perception that remastered releases are always 'brighter and louder' than the original releases.

The next measurement undertaken was LUFS, a loudness measurement similar to RMS in terms of calculating average loudness, but which takes into consideration human perception of audio loudness and is an interleaved measurement (not separated by left and right).

Table 4.3 represents LUFS measurements across both versions. Similarly, to the RMS readings, the original 1995 CD version of 'Wonderwall' is significantly quieter (-2.2dB) compared to the remaster, whereas 'Champagne Supernova' is louder (+0.9dB) than the remaster. This is inconsistent with the levels analysed for both *Abbey Road* and *Goodbye Yellow Brick Road*, where a clear pattern emerged of the remaster being consistently louder across all tracks than the original vinyl release. These results tend to support the use of hyper compression and/or brick wall limiting used on the original release in that there was not a lot of room available for volume manipulation when it came to producing the remaster.

Table 4.4 depicts decibel measurements of dynamic range (DR) across both releases of the three song recordings analysed. Again, we have inconsistencies across the two versions. The original 1995 release of 'Wonderwall' has the greatest DR reading of all tracks, and also has the greatest reduction in DR for the remaster, which is consistent with the results depicted in the reduction in DR on The Beatles and Elton John case

TABLE 4.3 LUFS level measurements *(What's the Story) Morning Glory?*

Song	*1995 mastered CD*	*2014 remastered CD*
	LUFS	*LUFS*
'Wonderwall'	-8.13	-5.9
'Don't Look Back in Anger'	-6.71	-6.25
'Champagne Supernova'	-5.42	-6.5

TABLE 4.4 Dynamic range measured in dB *(What's the Story) Morning Glory?*

Song	*1995 remastered CD*	*2014 remastered CD*
	Dynamic range	*Dynamic range*
'Wonderwall'	7	5
'Don't Look Back in Anger'	5	5
'Champagne Supernova'	3	4

studies, analysed previously. There is no change in DR for 'Don't Look Back in Anger'. However, the DR for 'Champagne Supernova' is increased in the remaster compared to the original, which is the complete opposite result you would expect based upon the trends previously established.

To further examine the characteristics of a reduction in DR and increased loudness inherent in the digital replicas, as opposed to the original analogue music artefact, the next measurement undertaken was frequency spectrum analysis. Similar to O'Malley's work, I was keen to examine the frequency spread of the different *(What's the Story) Morning Glory?* versions in an attempt to identify where these dynamic differences existed and why they were made (2015). It is important to note that the images in Figures 4.4–4.6 only provide a brief snapshot in time on the various frequency and volume levels across both releases.

Figure 4.4 represents the frequency spectrum for the song 'Wonderwall', the third track on the CD. The image was captured at around 55 seconds into the song in the second verse where the lyric 'I' at the start of 'I don't believe that anybody' is sung. The light grey colour represents the 1995 master and the dark grey colour is the 2015 remaster.

It is evident that the 2014 remaster is louder across all frequencies than the 1995 original release and the frequency spectrum shape is similar. However, there appears to be less difference in volume across the 2–3kHz range for the remaster which would sound like a slight cut in this range on the remaster overall. This lends support to Brine's perception that the vocals appeared to sit better on the original (this range would appear to have a slight boost on the original compared to the remaster).

Figure 4.5 represents the frequency spectrum for the song 'Don't Look Back in Anger', the fourth track on the CD. The image was captured at around 4 minutes and 40 seconds and is the last lyric sung, 'day', in the song. The light grey colour represents the 1995 master and the 2015 remaster is dark grey.

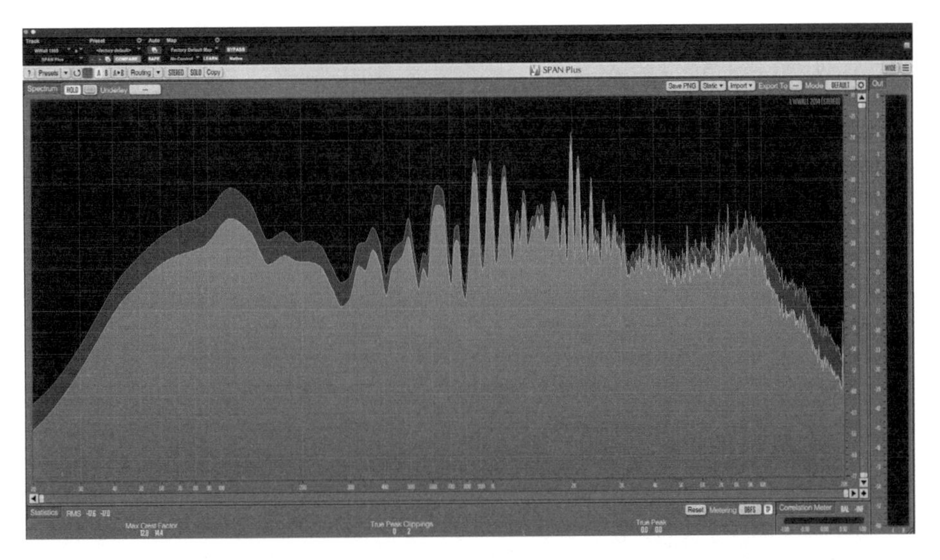

FIGURE 4.4 Spectrum analysis image for 'Wonderwall' – 1995 (light grey) and 2014 (dark grey)

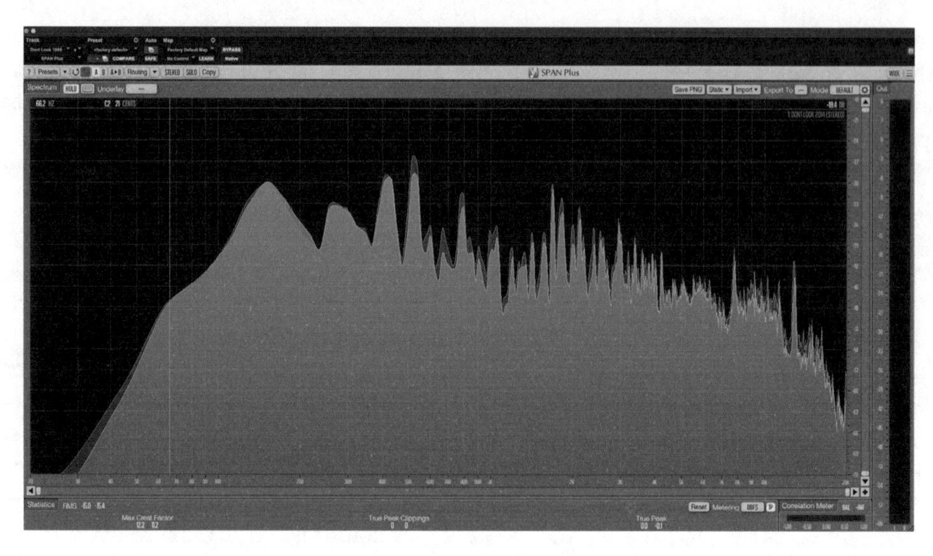

FIGURE 4.5 Spectrum analysis image for 'Don't Look Back in Anger' – 1995 (light grey) and 2014 (dark grey)

It is evident that the 2014 remaster has a slightly increased lower end (below 20Hz) and is louder across the spectrum from around 500Hz onwards. There also appears to be more energy on the remaster around 7–10kHz and 14–16kHz. This would suggest that there is greater sparkle and air on the remaster. The range 150–250Hz is louder on the original which may support Brine's observation that there were some point areas removed or cut in the remaster.

Figure 4.6 represents the frequency spectrum for the song 'Champagne Supernova', the last track of the CD. The image was captured at around 4 minutes and 29 seconds where the first 'why' is sung from the phrase 'I don't know why, why, why, why'. The light grey colour represents the 1995 remaster and the 2015 remaster is dark grey.

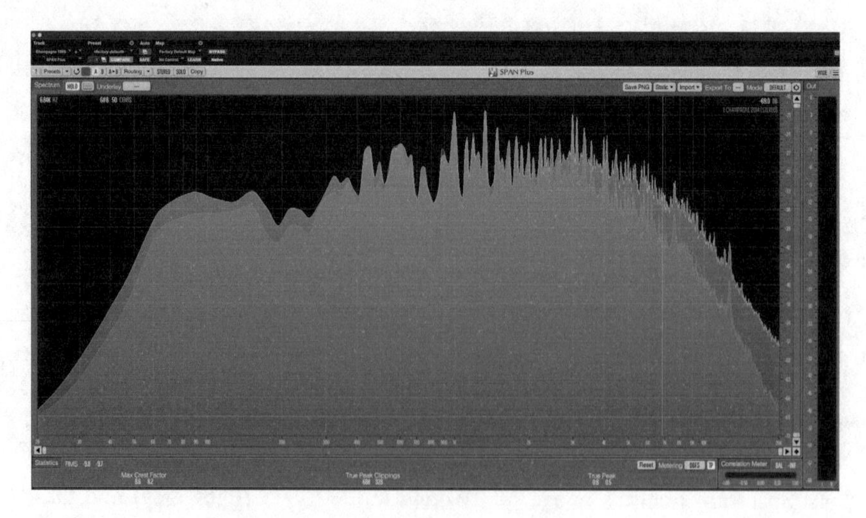

FIGURE 4.6 Spectrum analysis image for 'Champagne Supernova' – 1995 (light grey) and 2014 (dark grey)

This is perhaps the most surprising frequency spectrum image in the book, let alone this chapter. This is the only frequency spectrum image that clearly shows that apart from some small pockets within the range 500Hz–1.4kHz, the 1995 original is clearly louder across all other frequencies than the remaster. Furthermore, the shape suggests that there is greater emphasis in the ranges 20–400Hz and 4kHz–20kHz suggesting not only a stronger sounding lower end, but also much greater presence, sparkle and air. This lends support to Brine's perception that he liked 'the honky sound of the guitars on the original where the mid frequencies are stronger'. Additionally, the increase in the dynamic range of the remaster also supports the claim made by Brine that the remaster appeared to sound more 'open'.

Cultural heritage

Nick Brine

In general, I am not against remasters. I find repackaging and remastering iconic albums that include bonus tracks and alternate mixes that weren't available previously when the original version was released is good for both the fans and the bands themselves, particularly as a source of revenue, which can be much needed in some circumstances. I don't feel the same way about remixing however, unless it's about getting in someone specific to create a clearly different 'signature' sound and feel. Digital remasters open up a variety of formats for audiences which I also think works well. Furthermore, it is nice being able to use today's digital equipment to get rid of hiss or noise or some technical issue you could not remove previously or add more bottom end than was possible on vinyl, as this can massively enhance your listening experience.

I think being able to listen to something that will sound as good as it possibly can on the format available at the time, which will no doubt be different in 20 years' time, is a good thing. I mean those old Beatles' records and other classic recordings will always stand the test of time because the songs and performances are so good. And any future remasters of these classic songs should not vary significantly as these song recordings are set in stone. I think it would be sad if people got into the original files and started editing stuff and moving things around, remixing it and adding effects. However, I see this as a different activity that is never going to replace the original and believe people will always want to have access to the original version.

Ian Cooper

I would prefer remastering to be undertaken in the analogue chain as opposed to digitally. I have never mastered anything digitally in my life and I know of one or two recordings where the credits read 'analogue remastered by Ian Cooper'. If you have a proper tape machine, then go back and use the originals – that's how I believe remastering should be done. But record companies don't necessarily have the tapes, nor want to give them to you, so they may try and give you a CD copy from the 1980s instead. Does it really matter? There are an awful lot of hit songs out there with appalling sound. When I was at PYE, there was an engineer who had a Grammy for his production work, and I use to work with his producer. Anyway, his mixes always sounded lousy, so much so that one day he brought in a tape of his and I set up the Pultec EQs quite aggressively prior to listening to it, because I knew what it would sound like. Anyway, this award-winning engineer queried what I was doing so I played his version without my EQ and then my enhanced version. He could not believe the difference and thought all along what he was producing was great, when it was not.

It doesn't worry me if people remaster my records I have previously mastered. I don't listen to much music at home, in fact I couldn't think of anything worse. I know people who have redone stuff and said to me 'I'm working on one of your records and it sounded really good' which is nice. I don't think digital sounds particularly nice and think it is not quite there yet compared to analogue.

Bibliography

Barry, B. (2013). *(Re)releasing The Beatles.* Paper presented at the Audio Engineering Society Convention 135, New York, USA.

Buskin, R. (2012, November). Oasis 'Wonderwall'. *Sound on Sound*, accessed 11 June 2021, https://www.soundonsound.com/people/oasis-wonderwall.

Huggins, D. (2011). Owen Morris: How I mastered Morning Glory, accessed 11 June 2021, http://www.oasis-recordinginfo.co.uk/?page_id=6.

Oasis (2020a). Oasis – (What's The Story) Morning Glory? Return to Rockfield, online video, YouTube, accessed 11 June 2021, https://www.youtube.com/watch?v=Mg9JwZSG_vc.

Oasis (2020b). Oasis – (What's The Story) Morning Glory? Track by track with Noel Gallagher, online video, YouTube, accessed 11 June 2021, https://www.youtube.com/watch?v=lQISVjDAw7w.

Oasis (2020c). Rockfield Studios – Interview mix Noel & Liam Gallagher, Bonehead – Morning Glory 25 years, YouTube, accessed 11 June 2021, https://www.youtube.com/watch?v=8vpXgK58P0U.

Oasis Recording Information (2021). Equipment used in the Morning Glory sessions, accessed 11 June 2021, http://www.oasis-recordinginfo.co.uk/?page_id=393.

Official Charts (2020, 14 February). The best-selling singles of all time in the Official UK Chart, accessed 14 July 2022, https://www.officialcharts.com/chart-news/the-best-selling-singles-of-all-time-on-the-official-uk-chart__21298/.

O'Malley, M. (2015). The definitive edition (digitally remastered). *Journal on the Art of Record Production* (10).

Owsinski, B. (2008). *The Mastering Engineer's Handbook: The Audio Mastering Handbook.* Cengage Learning.

5

REMASTERING MOZART'S *THE MAGIC FLUTE*

Introduction

I included this case study to investigate remastering as applied to live and orchestral operatic recordings as opposed to solely concentrating on rock and roll and studio recordings. Mozart composed his famous two-act opera *The Magic Flute* in 1791, a few months before his death. At the time, the opera was controversial as it appeared to be embedded in the ideals of Freemasonry, a secret society, and in conflict with the ruling Austrian Hapsburgs. Shrouded in mystery, mythical creatures and magic, the opera is open to many interpretations, and in combination with Mozart's brilliant and powerful score, the opera is still popular today (Mozart, Fisher, Schikaneder and Giesecke, 2005).

The opera was performed and broadcast live on BBC Radio in 1962 from the Royal Opera House in Covent Garden, London. The Orchestra of the Royal Opera House and the Royal Opera Chorus were conducted by Otto Klemperer. This performance is quite rare as it features Dame Joan Sutherland, arguably one of the greatest operatic sopranos of all time, in her only known recording of the part of The Queen of the Night. Considered lost or incomplete, a recording of the full programme complete with BBC Radio announcements was discovered in a collection and believed to be 'originally prepared for or by one of the performers' (Rose, n.d.). This recording was subsequently remastered by Paul Baily in 2014 for the Testament label and was officially released on 13 April 2015.

Background (Paul Baily)

I always had an interest in music and as a child I learnt to play the piano and read scores to a reasonable standard. In fact, I used to track down miniature scores in music shops of classical recordings my parents had in their vinyl collection so I could follow along as I played the records. I left school and attended the University of Newcastle upon Tyne to undertake a music degree. Back then music degrees were mainly academic and based around classical music, but they did have an analogue studio that we

DOI: 10.4324/9781003177760-6

could access and that was of particular interest to me. Along with a couple of other students (no one else was interested) we used to spend hours in there playing around with REVOX reel-to-reel machines and making tape loops. It was great fun and we learnt so much from making plenty of mistakes. From there I moved nearby to Durham to complete a master's degree in composition. The university had a slightly larger studio including a 24-track homemade mixing desk.

Around this time, I volunteered at BBC Radio York which soon led to paid work. I got hold of a Music Yearbook and typed up and sent 200 letters of enquiry regarding work opportunities with studios and record labels. As you would expect, most companies did not respond, but I was invited to EMI offices in Gloucester Place, London, for an interview with Peter Alward, the then A&R Director of EMI Classics. During the interview, Peter said it was obvious I did not want to work there and that I may be more suited to working at Abbey Road Studios. He sent a memo to Ken Townsend, the then boss of Abbey Road Studios, who contacted me later and invited me in to have a look around. While there, I met the editor David Bell who had a tape lined up and a large score of one of the Stravinsky ballets. David then played the tape and pointed to a section of the score and said 'OK, press this button when you get to this part of the score'. So, I pressed the button at the correct time and was offered a job in the editing department!

When I joined EMI in 1986, they had only switched from analogue to digital editing six months prior. They had state of the art equipment at the time, including a Sony 1100 (the 3000 came later) digital editing system. Basically, you had two U-matic tape machines, one with the source material on it and the other acting as a destination file that you built up as the master. You would put about eight seconds of audio around the edit point and this would be stored in the buffer memory. You would press 'preview' to check the edit and hear it back, and then once happy with it, you would press 'record' and the destination machine would record the edit. You could only go up to around 30–40 millisecond crossfades so it could be quite challenging to make them sound seamless.

Around this time, Abbey Road Studios built a purpose room for remastering from analogue ¼-inch tape to digital which would be pressed to CDs, the new and popular format at the time, and I started working in that studio. We would run the signal from the source tape machine into a DDA mixing desk where we could apply outboard effects like EMI EQ (brilliant) and Urei 'little dippers' that remove hums and a Lexicon reverb to add a bit of life to old recordings. From there the signal went through the AD convertor and onto the destination U-matic tape. This was then sent up to the PQ department where a PQ master was made and then shipped off to the factory where a CD glass master was made from which CDs were pressed. Highlights from that time working for EMI include remastering most of the Maria Callas back catalogue, working with Nigel Kennedy on the Elgar *Violin Concerto* and Vaughan Williams *The Lark Ascending* and remastering the complete set of Beethoven Symphonies performed by the Philharmonia Orchestra and conducted by Herbert von Karajan.

I left Abbey Road in 2000 and set up my own studio called Re:Sound and built a room which was pretty much a carbon copy of my Abbey Road room, except I monitor through ATC100s which are superb. Of course, computers had come into play at this stage, and my workflow now included (and still does) the addition of a Pyramix DAW, which is great. I often need to do complex editing and Pyramix is one of the best

FIGURE 5.1 Paul Baily in Re:Sound Studio [Photograph], received from Re:Sound

DAWs for this (along with Sadie and Sequoia). One of the reasons I left Abbey Road was that staff tend to specialise in one area, whereas I was keen to branch out into recording and mixing which I have been able to do. My work at Re:Sound includes recording many orchestral artists, remastering the BBC Legends label as well as remastering for the Testament label.

Original production of *The Magic Flute* 1962 (Paul Baily)

Regarding the original production, it was made by the BBC for broadcast on The Third Programme, a station which featured classical music, before it was eventually replaced by Radio 3. It was recorded and broadcast in mono, which the BBC tended to focus on as almost all radios at the time were mono.

Obviously, I wasn't there so I can't really comment on microphone positions and levels etc., but common techniques at the time for orchestral recordings were the 'DECCA tree' and 'outriggers' to capture a full rich sound. As this is an opera, they may have had a few extra microphones set up for the soloists and chorus to use. They most likely used a selection of large diaphragm Neumann microphones. However, I know the BBC used a lot of Coles 4038 ribbon microphones at the time, but I'm not sure whether they would have used these in a live operatic recording. As an aside, when the BBC did begin stereo recording, they tended to record mid/side – the stereo signal was sent in the modulated part of the FM radio wave and where this was lost in poor reception there would still be the mono part to listen to.

There would have been two tape machines set up. This is because tape running at 15 inches per second (IPS) would last around 30 minutes, so you would need to have the second machine recording before the first machine ran out of tape. They would have then, later after the performance, edited it together to possibly be broadcast again at a later date.

Remastering *The Magic Flute* 2015 (Paul Baily)

I remastered *The Magic Flute* in 2014 in my studio and it was released in 2015. I received a copy of the original ¼-inch master tape from the BBC and played it out from my Studer A820 two-track tape machine. From there, the signal was sent through my dCS analogue to digital convertors into Pyramix. I used the effects available within Pyramix (EQ, reverb and compression) and CEDAR ReTouch to make changes and to remaster, although I do have an outboard Lexicon reverb unit I sometimes include in the chain. Once I completed the remaster, I bounced out a WAV master and sent it off to the label for feedback. I can't recall if they came back with any changes; generally this doesn't happen too often unless there are specific issues with the sound.

It is interesting to note that this particular recording has been remastered and released on Testament and also another label. My remaster for the Testament label was the later release and, to me, sounds better. When I remaster anything, I try to make sure mine sounds better than previous remasters done by others, as there is no point otherwise. The earlier version sounds flatter and there appears to have been a lot of width added to what would have most certainly been a mono recording in an attempt to make it sound more, for want of a better term, 'stereo'. I always try to be faithful to the original recording and if it is mono, I may (if required) widen it slightly to produce a richer and fuller sound, but certainly not to the extent of the earlier remastered version. Widening mono recordings can introduce many issues, most noticeably phasing, so you need to tread carefully in this space.

The general layout for an orchestra, from the audience's perspective, is violins on the left (treble) and bass on the right. That is a key difference to remastering pop and jazz albums which I have also worked on. For example, with a pop record you generally have the bass pumping down the centre along with kick and snare to really drive the song along. With orchestral recordings you have bass on one side and treble on the other so I try to balance this as best as I can without the sound being skewed towards one side or the other. As with recordings in most genres, the aim is to achieve a natural frequency balance across the spectrum from low to high frequencies.

When I first listen to old recordings I really try to focus on the EQ, in particular the frequency range of 2–3kHz, where the ear is most sensitive. A lot of the early EMI recordings can sound a bit harsh in this area with the associated harmonics flying around, due to being emphasised in the 2–3.5kHz range to compensate for poor quality playback machines of the day (at least by comparison to devices used today), so I generally start here and cut a few dB just to warm it up a little. To create a richer and fuller sound and better listening experience I may add some treble for the violins and some bass to help the lower instruments.

On this particular recording, I reduced the right side or bass track at around 5.6kHz and boosted at around 100Hz, and on the left side for the violins I boosted at 5.6kHz and cut at around 100Hz. These changes are only a couple of dB, nothing too dramatic. I didn't use a stereo width plug-in as such, as I find that my technique works better and sounds more natural without introducing phase issues. Early recordings often contain a lot of rumble and unwanted low frequencies. To avoid removing any bass content, I always automate a high pass filter which I use where necessary. It helps to be reading the score to know when the basses are about to start!

I may use parallel compressions to give a recording a bit more body through bringing up the quieter material as opposed to squashing a recording down from the top. I sometimes use a brickwall limiter but only to raise around 2dB in gain with a ceiling at around -0.7dB. I feel that much care should be exercised if using compression because this alters the internal frequency balance. Always listen carefully to check that the sound has been enhanced and not degraded.

Regarding hums and whistles and crowd noise that often occur with live orchestral and operatic recordings, I generally listen through on headphones to identify them. I find headphones make noise and issues more noticeable and easier to narrow down. I will also mark up a session at this stage, making little notes of what needs to be fixed where. If it is an electrical click, for example which is common for tape that has been edited together, I use ReTouch to simply draw around the offending transient and remove it. If an audience member coughs during the performance, which is probably more often than you would realise, then that is a far more complex operation. A cough has a duration and decay attached to it across a particular frequency range, so the aim is generally to reduce as opposed to remove it, as it can cut through the frequency range of the music. Again, using ReTouch, I can get in between the harmonics to reduce the cough while minimising the effect on the music. Audience members also tend to cough in between musical segments if they can, as a means of courtesy to others most likely, but this can still cause problems. It is because the music generally decays from one movement through to the next so a simple cross fade will affect the music. In this case I would generally copy and paste a bit of ambience from another section, smoothing it into place and adding a tiny bit of reverb, which tends to work well.

For *The Magic Flute*, I worked solo, without the client in the room, which is my preference. I have found through past experience that having a client in the room can make the process three times longer and you risk going around in circles listening to different versions without making any firm decisions. I also generally work with and follow the score of the piece I am working on to make sure I am delivering an accurate representation of the work.

Comparative listening (Paul Baily)

The main differences I hear between the two versions is that the remaster has had the electrical mains hum and rumbles filtered out as well as noise removed. The removed noise includes electrical clicks and sounds from the audience, although it is not always possible or even desirable to remove all audience noise.

The remaster has been warmed up through EQ and you can hear that the top end is less brittle and that there is more focus on the bass. I can also hear the small amount of medium hall reverb I applied and as it is a mono recording it is not so noticeable. There is a slight width added to the remaster (I'm not a fan of this, but the client wanted it) and parallel compression for more body and to make the recording sound fuller.

Digital audio analysis

Paul Baily sent me a digitised copy of the flat transfer taken from the original ¼ inch tape recording of the BBC Radio live performance of Mozart's *The Magic Flute* in 1962 from the Royal Opera House in Covent Garden, London, alongside the 2015 remaster which were both WAV files at 16bit/44.1kHz WAV files. I decided to focus on approximately the first 13 minutes of the opera as this provides contrast between the orchestral 'Overture' followed by 'Have mercy, have mercy' from Act 1 which features vocals. I implemented the digital audio analysis elements outlined in the methodology.

Figure 5.2 is a screenshot that depicts the loudness/amplitude range over time of the respective waveforms for both versions of *The Magic Flute* within the Pro Tools DAW software environment. The top horizontal track represents the digitised flat transfer, and the bottom track depicts the remaster. Despite containing two separate elements, the 'Overture' and 'Have mercy, have mercy', I have displayed each WAV files as whole (one file containing both elements) to include the applause from the audience during the break between musical pieces. Although there does not appear to be significant differences in amplitude between the two versions, we can clearly see that the remaster has a shorter duration, and this is due to pitch. As Paul Baily explains:

> With all recordings (especially older ones) I always check the pitch, so that A = 440. This is crucial with 78rpm records, but back in the 50s and 60s the electrical voltage supply wasn't always consistent and tape machines could sometimes run slightly fast or slow. The Berlin Philharmonic and Vienna Philharmonic play slightly sharper so never assume anything! I then calculate the percentage difference to correct the pitch and do a sample rate conversion, creating a new file at the correct pitch.

FIGURE 5.2 Waveform view of digitised 1962 flat transfer and 2015 remastered versions of recordings from *The Magic Flute*

Although there appears to be no major variances in amplitude between the two versions based upon the waveform image, I measured the left and right true peak levels to explore this further. I decided at this stage to split the 'Overture' and 'Have mercy, have mercy' into separate tracks so that analysis could be undertaken on both orchestral and vocal pieces separately to determine potential differences between them.

Table 5.1 represents the left and right true peak level values for the two selected musical pieces across both versions. It is clear that there is minimal difference in peak level readings for the 'Overture' when comparing the digital flat transfer to the remaster. However, for 'Have mercy, have mercy' the remaster has its loudest peak at approximately 0.6dB lower than the digitised flat transfer.

Although this is helpful information in terms of identifying the loudest peaks across the two musical pieces, RMS measurements may be more helpful in determining any potential differences in 'average loudness'.

Table 5.2 displays the RMS levels for both left and right channels across both versions. It is evident that the 'Overture' remaster is approximately 2dB louder than the digitised flat transfer. However, both versions of 'Have mercy, have mercy' are fairly similar in average loudness, with the remaster showing a slightly louder RMS measurement.

To further explore potential variances between both versions in average loudness, LUFS also consider human perception of loudness and the measurements are depicted in Table 5.3.

Table 5.3 represents LUFS measurements across both versions. It depicts the remastered version as approximately 0.5dB 'louder' than the digitised flat transfer for the 'Overture', and that it is about 0.5dB 'quieter' for 'Have mercy, have mercy'. It is interesting to note that the 'Overture' is approximately 4–5dB louder than 'Have mercy, have mercy' across both versions which is most likely an indication of the full orchestra playing loudly in parts, particularly where vocals are not included.

TABLE 5.1 True Peak level measurements *The Magic Flute*

Song	*1962 digitised flat transfer*		*2015 remastered CD*	
	Left true peak	*Right true peak*	*Left true peak*	*Right true peak*
'Overture'	-0.47	-0.46	-0.46	-0.47
'Have mercy, have mercy'	-0.46	-0.46	-0.51	-1.11

TABLE 5.2 RMS level measurements *The Magic Flute*

Song	*1962 digitised flat transfer*		*2015 remastered CD*	
	Left RMS	*Right RMS*	*Left RMS*	*Right RMS*
'Overture'	-16.61	-16.61	-14.70	-14.88
'Have mercy, have mercy'	-19.84	-19.84	-19.15	-18.99

TABLE 5.3 LUFS level measurements *The Magic Flute*

Song	*1962 digitised flat transfer*	*2015 remastered CD*
	LUFS	*LUFS*
'Overture'	-14.4	-13.90
'Have mercy, have mercy'	-18.26	-18.78

Table 5.4 depicts decibel measurements of dynamic range (DR) across both versions of the two musical pieces. It is evident that the DR of the digitised flat transfer is greater than the remaster for both pieces, although only slightly, suggesting that in order to make the remaster 'louder' that there was some lowering of the DR perhaps through compression and/or limiting. Baily said that he added parallel compression for more body and this may have contributed to the lower DR score on the remaster.

To explore these differences in DR and average loudness, the next stage of the analysis was to use frequency spectrum analysers to determine potential variances across various frequencies. Although the frequency spectrum only displays a snapshot in time, it is still quite useful in determining variances.

Figure 5.3 represents the frequency spectrum for the 'Overture' and occurs at approximately 1 minute and 41 seconds. In order to be as accurate as possible, I shifted and realigned the remaster waveform with the digitised flat transfer to compensate for the duration differences due to pitch. The light grey colour represents the digitised flat transfer and dark grey is the remaster.

It is evident that the remaster has more energy and amplitude across the whole frequency range except for around 2–4kHz where the digitised flat transfer is louder. This appears consistent with Baily's notion that older recordings tended to have a lot of presence in this range to compensate for playback devices at the time and that he generally reduces EQ in this range. It is also supports Baily's tendency to warm up the lower end of older recordings by increasing bass, particularly 50–120Hz, and to add some air and sparkle from 4k to 18kHz. There is also a spike evident on 20kHz which may have to do with the electrics associated with the original recording.

Figure 5.4 represents the frequency spectrum for 'Have mercy, have mercy' and occurs at approximately 1 minute and 8 seconds into this musical work. Again, I shifted and realigned the remaster waveform with the digitised flat transfer to compensate for the duration differences due to pitch. The light grey colour represents the digitised flat transfer and dark grey is the remaster.

TABLE 5.4 Dynamic range measured in dB *The Magic Flute*

Song	*1962 digitised flat transfer*	*2015 remastered CD*
	Dynamic range	*Dynamic range*
'Overture'	12	10
'Have mercy, have mercy'	14	13

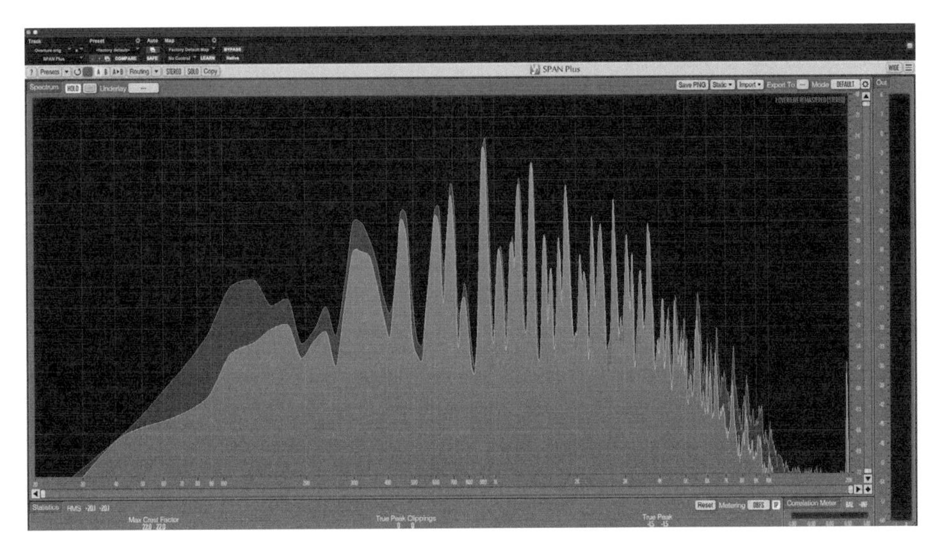

FIGURE 5.3 Spectrum analysis image for *The Magic Flute* 'Overture' – 1962 (light grey) and 2015 (dark grey)

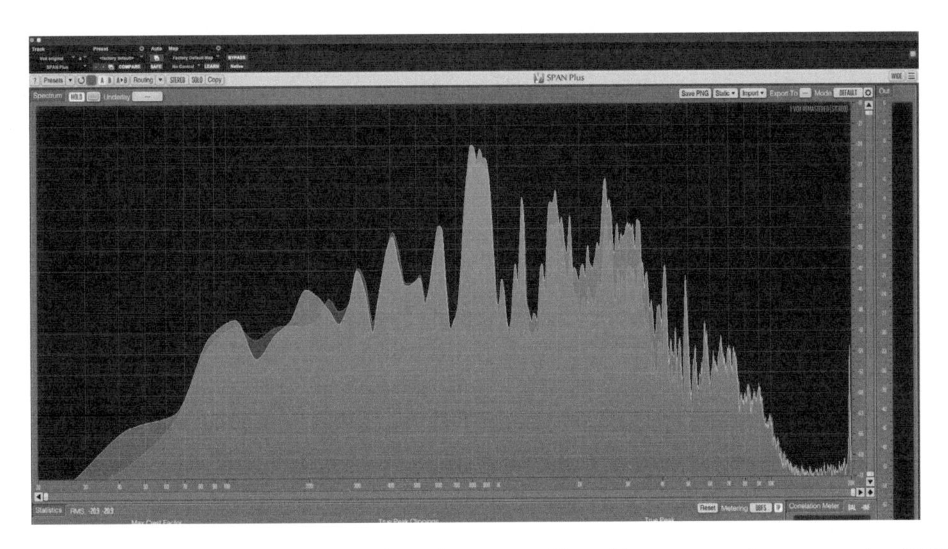

FIGURE 5.4 Spectrum analysis image for *The Magic Flute* 'Have mercy, have mercy' – 1962 (light grey) and 2015 (dark grey)

It is evident that the remaster only displays slightly more amplitude in frequency notches including 140Hz and 250Hz. The remaster also appears to display major cuts between 1.5–5kHz which is consistent with Baily's approach to warm up the bass and reduce presence. Additionally, the remaster has much less energy 20–70Hz which is consistent with Baily's approach to using high pass filters to remove rumble. Again, here is also a spike evident on 20kHz which may have to do with the electrics associated with the original recording.

Cultural heritage (Paul Baily)

I believe there are two schools of thought on whether it is culturally appropriate to digitally manipulate older recordings. Personally, I believe it is OK because the bottom line is that I aim to present the music to the public in a way in which they are going to enjoy it and want to listen to it more than once. And if that means manipulating it in some way, then I will, but only so much. For example, *The Magic Flute* has a little bit of stereo widening added to it (I must have been asked to do this), so I tried to do this in the most least offensive way through adding some treble to the bass side and some bass to the treble side as opposed to using a stereo widening plug-in with all the associated phasing problems that come with it.

I had another client who was very keen to have all of these old mono recordings remastered and widened to sound stereo, which to me is not the right approach. These are mono recordings, not stereo. I appreciate that this can be done through plug-ins and tampering to make it appear that the sound is being pushed out towards the sides, but it is phoney and not something I support.

With regards to remastered versions potentially replacing the original productions as the authentic version for future generations, I feel this applies to studio recordings only as opposed to recorded live performances. I think it is important to note that when music is first released to the public, say a Beatles' record on vinyl, that this will sound different to what the master tape in the studio sounds like. This is due perhaps to the cutting engineer perhaps rolling off the bass, other compromises associated with using the vinyl format, limitations of the consumer playback device and the degradation that occurs to vinyl after numerous plays. So, you could argue that the public never actually get to hear the original production as it actually sounded in the studio.

I am the chairman of a non-profit organisation called Music Preserved, which aims to digitise and rescue historic recordings of rare and unreleased material including radio broadcasts. I have come across people who believe that digital replicas of these recordings should be strictly flat, warts and all, with no digital manipulation whatsoever. From a scholarly viewpoint this is interesting as you can get a sense of what it would have been like to listen to a 1942 radio broadcast in 1942, but from a listener's perspective whereby the goal is to make the listening activity as pleasurable as possible, and in line with current live recordings of orchestral works, digital manipulation is definitely required.

Bibliography

Mozart, W.A., Fisher, B.D., Schikaneder, E., & Giesecke, C.L. (2005). *Mozart's The Magic Flute.* Opera Journeys Publishing.

Rose, A. (n.d.). KLEMPERER Mozart: Die Zauberflöte, Symphony No. 41 (1962/1954) – PACO115. *Pristine Classical*, accessed 12 September 2022, https://www.pristineclassical.com/products/paco115?_pos=1&_sid=60eb30419&_ss=r.

CONCLUSION

This book has revealed that technology and digital tools have transformed the role of mastering from a once purely technical and compliant role into a creative and powerful one. Traditional mastering or cutting engineers were restricted in what they could do to uphold the quality of the final master tape mixes onto the vinyl format, given technology available at the time as well as physical limitations of the media. This would often result in the vinyl version sounding 'inferior' compared to the master tapes due to the constant compromises cutting engineers were forced to make when transferring from tape to vinyl, as well as inherent noise of the vinyl format during playback. Current mastering engineers however (unless they are mastering to vinyl) do not have the same restrictions mastering to a digital format. Therefore, with these vinyl format limitations removed, and the vast array of digital tools available, modern releases can vary greatly sonically between mix and master stage, and as the final stage in the music production workflow, the decision-making process of the mastering engineer is an important and significant one.

If there can be great variance between modern mixes and masters due to technology then it would suggest that remastering older recordings could potentially lend itself to even greater variance, particularly around perceived increases in loudness and bass. The 2009 remastered version of The Beatles' *Abbey Road*, and both the 1995 and 2014 remastered CDs of Elton John's *Goodbye Yellow Brick Road*, emanating from professional studio analogue recordings and technology from the late 1960s and early 1970s, appears to support this view. For example, the data showed significant increases in LUFS readings for these remasters of roughly 5–6dB compared to the flat digitised version of the original production. Additionally, the frequency spectrum analysis revealed a general trend for increased bass (40–200Hz) across the remastered versions of these albums. This trend of increased perceived loudness and bass is consistent with results from previous research undertaken (Bruel, 2021a; Bruel, 2021b) on Australian band Sunnyboys' 2014 remaster (professional studio recording from 1981) and the 2016 remasters for Jumble Sale (home studio recording from 1986) and Ben's Calf (semi-professional studio recording from 1995).

DOI: 10.4324/9781003177760-7

However, to conclude that remastering practice predominantly revolves around perceived increases in bass and loudness is naive. In the case of Oasis, although their iconic album *(What's the Story) Morning Glory?* was recorded in 1995 using a professional studio workflow resulting in an analogue stereo master tape, at a time when DAT was burgeoning as the preferred media, it was mastered and released on both vinyl and CD. Being mastered for release on CD removed the physical limitations of vinyl resulting in a much louder version than possible on vinyl. In fact, this 1995 CD release is credited as a major contributor and pioneer of the so called 'loudness wars' (Hjortkjaer & Walther-Hansen, 2014; Nielsen & Lund, 1999; Vickers, 2010; Walsh, Stein, & Jot, 2011).

The Oasis case study revealed that the 2014 remaster of *(What's the Story) Morning Glory?* resulted in a lower (although slightly) LUFS reading for the track 'Champagne Supernova' compared to the 1995 original version. This lowering of perceived loudness for the remaster is most likely the result of the original being recorded and mastered at such a high level. If we compare LUFS readings across the case studies, it is clearly visible that the 1995 Oasis CD release is around 10db louder than the 1969 The Beatles' *Abbey Road* vinyl digital transfer and approximately 9dB louder than the 1973 Elton John's *Goodbye Yellow Brick Road* vinyl digital transfer, which is quite significant. The main sonic differences between the original and remastered Oasis album therefore appears to be more focussed on slight EQ changes as opposed to overall perceived loudness.

It is, however, far too simplistic to adopt an assumption that the phenonium of the remastered version being 'less loud' than the original release is restricted to the 'loudness wars', Brit Pop genre or technology and trends prominent at the time of recording. The Mozart case study revealed a similar result whereby the 2015 remastered version of 'Have mercy, have mercy' had a slightly lower LUFS reading, and therefore lower perceived loudness, compared to the digitised transfer of the 1962 recording. Similar to the Oasis case study, there were inconsistencies in perceived loudness (the remastered 'Overture' was slightly louder than the original) between the tracks. This may have to do with the genre of operatic and orchestral music where dynamics can vary greatly within the one music work.

Another general view of remastering music which requires further investigation is that remastered versions generally get louder over time with each reincarnation. The Beatles' case study certainly supports this view in that the 1987 CD remaster release was around 2–3dB louder than the original vinyl, and the 2009 remaster was approximately 2–4dB louder than the 1987 remaster. With a 22-year difference between remastered versions of *Abbey Road*, perhaps this difference in loudness is technological based, although it is documented that the 1987 CD remaster was basically the vinyl master printed on a CD with very little intervention as there was no apparent need (Bruel, 2019). The results of the Elton John case study, however, challenge this notion as the results depict that the 2014 CD remaster was slightly 'less loud' across all tracks analysed compared to the 1995 CD remaster.

There is a popular opinion in remastering that increasing loudness automatically results in a reduced dynamic range (a flattening out of the difference between the loud and soft parts) and that it is a necessary by-product of flattening and then increasing a signal. The Beatles' case study results revealed that although this assumption held true with most tracks, the 2009 remaster of 'Octopus's Garden' was approximately 3.5dB louder than the 1987 remaster but had a greater, although slightly, dynamic range score. Of course, this is again a comparison of the remastered versions only with no reference to the original release. If we include the data from the 1969 digital transfer, this version has the greatest dynamic range scores overall (although not significant). Similarly, the 1973 digital transfer of the original release of *Goodbye Yellow Brick Road* yielded the highest dynamic range scores, something of which David Hentschel is immensely proud. Interestingly, in the Oasis case study, the remastered version of 'Champagne Supernova' was less loud than the original, and had a larger dynamic range score, which shows that decreasing loudness can lead to an increase in dynamic range. However, the Mozart analysis revealed that the 2015 remastered CD version of 'Have mercy, have mercy' was less loud than the digitised flat transfer but also had a lower dynamic range score which questions this logical notion of increased perceived loudness automatically resulting in less dynamic range. Of course, this Mozart track was the only one across this book and in previous research (Bruel, 2019; Bruel, 2021a; Bruel, 2021b) I have undertaken to yield a result of less loudness resulting in a decrease in dynamic range, so it is most likely an anomaly of the genre and/or recording.

Regarding the 'loudness' measures of completed and released remasters, the research revealed that there does not appear to be an industry standard of perceived loudness, and that the loudness of remasters can vary quite significantly. For example, the LUFS measurements for the tracks analysed for the Mozart 2015 remastered CD were approximately between -18 to -13dB, The Beatles 2009 remaster was next loudest at -13 to -12dB, the Elton John 1995 remaster was similar to The Beatles at -12 to -10db, and the Oasis 2014 CD remaster by far the loudest at -7 to -6dB. It is interesting to note that in the case of The Beatles, Elton John and Oasis who are all iconic contemporary music artists who possess incredibly successful albums, even at this summit of commercial success with associated large budgets to utilise remastering professionals and studios, there appears no consistent or set industry standard of perceived loudness sought by the professional practitioners and record companies. I would expect to find these types of inconsistencies in perceived loudness regarding remastered releases in a sample of smaller independent self-financed and self-remastered releases undertaken on inferior equipment without the experience of professional practitioners (I may be wrong, and this may form an interesting study), so this revelation was surprising.

The case studies revealed that although we live in an age where we are surrounded by and immersed in digital technology, there is still room for analogue equipment as part of the modern remastering workflow, and also that each remastering engineer interviewed had their own specific customised workflow they implemented. Sam Okell described the workflow of remastering The Beatles as consisting of transferring analogue tapes to digital, noise reduction in the digital domain, then adding analogue EQs and then finishing with digital limiting. We can see the process constantly intertwines between digital and analogue. Tony Cousins detailed his workflow as coming off the analogue tapes into Dolby noise reduction then into a customised analogue mastering console, then through a digital convertor where digital limiting was applied. From there, the signal went to a digital console for Gus Dudgeon to make slight changes and finally into a DAW. Although a different workflow, it still evidences both digital and analogue stages. For the Oasis remaster, Ian Cooper used an entirely analogue workflow except for the digital conversion at the end into Sadie. In contrast, Paul Baily used an entire 'in the box' digital workflow (apart from transferring from analogue tapes) utilising Pyramix's native effects and CEDAR ReTouch for noise reduction and restoration to remaster *The Magic Flute* for CD. It appears therefore, that there is no one industry standard or workflow for remastering and that remastering practice varies greatly between the individual professional practitioner.

Another element of remastering practice that was evident in the research was the potential different final remastered product that would be achieved if a producer and/or artist of the original production was present during the remastering sessions. The Elton John case study revealed that remastering engineer Tony Cousins believed his 1995 remaster would have sounded different if he was left to his own devices as opposed to working alongside original producer Gus Dudgeon who Cousins described as 'focussed on trying to fix things' he was unhappy with from the original production. This also led, according to Cousins, to a longer timeframe to complete the project than what would have been if he was working solo although he still believes the 1995 CD remaster sounds fantastic. This intervention by the producer may be a contributing factor in the perceived sonic differences between Cousin's 1995 remastered CD and the 2014 remastered CD by Bob Ludwig, who worked solo on his project. In contrast, Ian Cooper describes working on the Oasis remasters alongside original producer Owen Morris (and sometimes artist Noel Gallagher) as a quick and easy process with minimal interaction and changes, as the original masters were so good, and they managed to remaster one Oasis album per day.

My experience of being an artist/producer present in the remastering process alongside a professional practitioner is documented in my previous research (Bruel, 2019; Bruel, 2021b). While there were some similarities with Cousins' description in that I felt I wanted to 'fix things' I was unhappy with regarding the original recordings, I also had tremendous faith and trust in the remastering engineer I was working alongside given his experience and standing within the music mastering and remastering industry within Australia. Furthermore, I believe I added value to and sped up the process in that I was able to quickly identify which versions of songs the remasters would emanate from, and my music production experience and background allowed me to input ideas and provide instant feedback when requested, during playback. On a subsequent remastering session regarding a different selection of recordings where I was not present, there were errors made concerning song versions which needed to be addressed, which made the process longer.

The research also revealed that there appears to be a clear delineation between audio preservation and remastering practice. Will Prentice from the British Library sound archive was adamant that his team was tasked to solely rescue and digitise the Library's collection in an 'objective' manner before these recordings could degrade further. This was based upon prioritising the rarity of the recordings and the degradation of the physical media housing them including wax cylinders, shellac discs, acetates/lacquers, vinyl and tape for digitisation. The team at the British Library sound archive refer to remastering as a 'subjective' activity that occurs at the next stage whereby researchers and archivists work on the digital transfers carried out by the Library. However, Mike Dutton detailed how his remastering practice concerning vintage media includes various techniques during and after the digital transfer process to achieve optimal results. The Beatles' case study depicted each team member of the 2009 remasters project as having a specific role and place in the remastering process. For example, Sam Okell described Simon Gibson's work of being mainly focussed on 'audio restoration' and that the process went from digital transfer to intensive listening to 'audio restoration' and then onto 'remastering' where EQ, compression and stereo widening may be applied.

There was a general view across the case studies that perceptions of cultural heritage and authenticity can be maintained through producing a digital replica remaster, where minimal intervention is implemented and whereby the context and meaning of the original production is maintained. For example, Sam Okell mentioned that remastering The Beatles in 2009, described at the time by project leader Allen Rouse as akin to 'fiddling with the Crown Jewels' (Sexton, 2009), had to be approached sensitively, with a clear plan, and constantly questioning that if the band and the engineers had the technology available at the time of the original production to remove or enhance certain sounds, would they? Furthermore, in the case study of the *Goodbye Yellow Brick Road* 1995 CD remaster, it is evident that producer Gus Dudgeon was 'actively' trying to change and/or fix things he was unhappy with in the original production, and a contributing factor to these perceived 'errors' may well have been technological limitations of the time.

However, there were some inconsistencies to this cultural heritage and authenticity mantra portrayed by the remastering engineers to always 'respect' the original production and intent of the artist, to enhance but not change. For example, The Beatles' 2009 remastering project of their official studio albums back catalogue (including *Abbey Road*) took a fairly large team of experts four years to complete, and the process consistently referred to the original productions throughout as a benchmark, to help guide what would and would not be regarded as 'acceptable' by the remaining band members, the team and the public. By contrast, the 1987 digital remasters release, undertaken by George Martin who also produced the original productions, included remixed (as opposed to purely remastered) versions of *Help* and *Rubber Soul* with noticeable added reverb which Martin thought would make them sound more 'modern'. These remixed versions actually replaced (they were not an add-on or extra version available within a package) the original mixed and mastered releases that had been adored by millions of people worldwide since their 1960s release. This inaccurate digital replica from 1987 actually replaced the original music artefact. This approach by Martin appears directly in contrast to a sense of respecting the context of the original recording which was produced by him!

For David Hentschel, digital remastering, however it is attempted and whatever approach is taken by the remastering engineers, is a violation of the original recording and the people involved in that. Hentschel views remastering as destroying a creative moment in history, a coming together of artists and professional audio engineers during a specific time period to produce these iconic music artefacts and works of art with the technology they had at the time. It could be argued that limitations in technology (limited tracks and no undo button) and expensive studio time (The Beatles used to record two to three songs in three-hour blocks) forced artists and engineers making analogue recordings to focus more on the quality of the 'performance', songwriting 'craft', and final mix and audio balance which ultimately led to these 'iconic' productions being created. For these iconic recordings to be digitally manipulated at a later date to have them sit more comfortably sonically (extra loudness and bass) against modern recordings, according to Hentschel, is ridiculous and culturally misrepresents the original music artefact.

The idea that remasters will potentially replace the original production over time as the authentic version, in that future generations will accept and view the remaster as the 'original' production through either a lack of access to the original media (not owning a record player and/or the original version not available on streaming platforms), is a very real proposition worth mentioning. Sam Okell is of the belief that the public will always have access to the original versions of The Beatles due to their ongoing popularity so that any remaster will be seen for what it is. Furthermore, the recent resurgence in vinyl production and record player sales implies a sense of optimism that vinyl media has a future and that original vinyl releases still have a solid place in our cultural landscape. However, currently digital streaming is by far the most popular means for the public to acquire and listen to music and there is no guarantee that vinyl production and playback devices will be sustainable in the long term. So where does that leave the cultural future of original productions and their sonic representations?

During my interview with Mike Dutton, we also briefly discussed the *Seventeen Seconds* album by British band The Cure he engineered in 1980. Dutton looked up the album on Spotify and the only version of this album available to the public is the remastered one which by its very nature is a digital replication, and more than likely with extra loudness, bass and possibly other sonic elements as is 'generally' the case with recordings from this vintage (Bruel, 2021a). There was no original production with its sonic characteristics available at all. If the original production is not accessible, then it is fair and, in some ways, logical to surmise that over time the remastered version of *Seventeen Seconds* will ultimately replace the original artefact as the 'authentic' version for future generations, and this is likely to occur to thousands of other original releases as well.

Remastering practice, it appears, is therefore not purely a technical and/or creative process which perhaps is a more suitable definition for the other music production elements, namely recording, mixing and mastering. While there exist similarities between remastering and these other studio-based production elements, in that a variety of digital and analogue tools and workflows are used in an attempt to create a sonically pleasing and professional final product to be distributed to the masses, remastering practice differs in that it has the added 'step' of cultural heritage considerations that need to be addressed. Unlike working on new musical works set to be released, where the audience has not heard the music before, or at least a professional master recording of it, the remastering engineer has to contend with, especially when working on iconic recordings, the knowledge that the public is most likely well versed in and familiar with the 'sound' of the original releases, and that there subsequently exists a pre-existing cultural and personal heritage-based relationship between the original recording and the public. With this in mind, it is therefore critical that current and future research into, and conversations around remastering practice, includes reference to its corresponding relationship with and potential impact upon cultural heritage.

Bibliography

Bruel, S. (2019). Nostalgia, authenticity and the culture and practice of remastering music. (Doctoral dissertation). Retrieved from https://eprints.qut.edu.au/129568/.

Bruel, S. (2021a). Remastering Sunnyboys. In, J. P. Braddock, R. Hepworth-Sawyer, J. Hodgson, M. Shelvock & R. Toulson (Eds.), *Mastering in Music* (pp. 155–173). Routledge.

Bruel, S. (2021b). Remastering the independent past. In V. Sarafian (Ed.), *The Road to Independence. The Independent Record Industry in Transition* (pp. 51–97). University of Toulouse 1 Capitol.

Hjortkjaer, J., & Walther-Hansen, M. (2014) Perceptual effects of dynamic range compression in popular music recordings. *Journal of the Audio Engineering Society*, 62(1/2), 37–41.

Nielsen, S. H., & Lund, T. (1999). *Level Control in Digital Mastering*. Paper presented at the Audio Engineering Society Convention 107, New York, USA.

Sexton, P. (2009, 12 September). Repaving 'Abbey Road'. *Billboard*, 121, 24.

Vickers, E. (2010). *The Loudness War: Background, Speculation, and Recommendations*. Paper presented at the Audio Engineering Society Convention 129, San Francisco, California, USA.

Walsh, M., Stein, E., & Jot, J.-M. (2011). *Adaptive Dynamics Enhancement*. Paper presented at the Audio Engineering Society Convention 130, London, UK.

INDEX